Energy conservation in textile industry

Energy conservation in textile industry

S. C. Bhatia (Author)

BE (Chemical), MBA

Prof. Puneet Mangla (Co-Author)

B.E. (Industrial Production), M.Tech. (Engineering Systems),
Head and Associate Professor, Department of Mechanical Engineering,
Hindustan College of Science and Technology, (Mathura – UP)

Edited by

Sarvesh Devraj

B.Tech (Mechanical), UPTU
M.Tech (Renewable Energy Engineering and Management), TERI University,
(Research Associate – TERI, New Delhi)

Published by Woodhead Publishing India Pvt. Ltd.
Woodhead Publishing India Pvt. Ltd.,
303, Vardaan House, 7/28, Ansari Road,
Daryaganj, New Delhi - 110002, India
www.woodheadpublishingindia.com

First published 2020, Woodhead Publishing India Pvt. Ltd.
© Woodhead Publishing India Pvt. Ltd., 2020

Woodhead Publishing India Pvt. Ltd. ISBN: 978-93-85059-39-1
Woodhead Publishing India Pvt. Ltd. e-ISBN: 978-93-85059-94-0

Typeset by Asian Enterprises, New Delhi
Printed and bound in India by Replika Press Pvt. Ltd.

Contents

Preface

The textile industry is one of the most complicated manufacturing industries because it is a fragmented and heterogeneous sector dominated by small and medium enterprises (SMEs). Characterising the textile manufacturing industry is complex because of the wide variety of substrates, processes, machinery and components used and finishing steps undertaken. Different types of fibres or yarns, methods of fabric production and finishing processes (preparation, printing, dyeing, chemical/mechanical finishing and coating), all inter-relate in producing a finished fabric.

Energy conservation is an important tool to deal with global issues such as the future exhaustion of resources and global warming. The energy supply chain begins with electricity, steam, natural gas, coal and other fuels supplied to a manufacturing plant from off-site power plants, gas companies and fuel distributors. Energy then flows to either a central energy generation utility system or is distributed immediately for direct use. Energy is then processed using a variety of highly energy-intensive systems, including steam, process heating and motor-driven equipment such as compressed air, pumps and fans.

Energy conservation contributes to solution to the global issues such as energy security and possible future exhaustion of oil. Industrial energy efficiency is a key ingredient in any national energy efficiency programme. Through the implementation of energy conservation, we can reduce the expenses for wasteful energy consumption and income will increase equivalent to the amount of the reduction. In order to promote energy conservation, it is effective to establish an energy saving activities framework. At the same time, it is also important to ensure a change in the attitude of energy consumers and to promote voluntary activities of energy conservation through performing the activities of publicity, awareness and dissemination of energy conservation. Textile industry uses large quantities of electricity and fuels. There are significant losses of energy within various operations of textile plants such as spinning, weaving and dyeing.

This book on Energy conservation in textile industry summarises various aspects of energy consumption and conservation and is divided into 22 chapters.

Chapter 1 deals with textile industry: An overview. The textile industry is not a single entity but encompasses a range of industrial units which use a variety of natural and synthetic fibres to produce various fabrics.

Chapter 2 is devoted to energy conservation in textile industry. In the textile industry, appreciable amounts of energy could be saved or conserved by regulating the temperature in the stem pipes, adjusting the air/fuel ratio in the boilers and installing heat exchangers using warm wastewater. Chapter 3 focuses on energy conservation in spinning process. Electricity is the major type of energy used in spinning plants, especially in cotton spinning systems. Various energy saving measures in spinning process have been discussed.

Chapter 4 concentrates on energy conservation in weaving process. Weaving machines (looms) account for about 50–60% of total energy consumption in a weaving plant. Humidification, compressor and lighting accounts for the rest of the energy used, depending on the types of the looms and wet insertion techniques. Various energy efficiency measures have been discussed.

Chapter 5 focuses on energy conservation in wet processing. Wet-processing is the major energy consumer in the textile industry because it uses a high amount of thermal energy in the forms of both steam and heat. The energy used in wet-processing depends on various factors such as the form of the product being processed, the machine type, the specific process type, the state of the final product, etc. Various energy conservation measures have been discussed. Chapter 6 deals with energy conservation in drying process. Drying is one of the most energy-intensive unit operations due to the high latent heat of vapourisation and the inherent inefficiency of using hot air as the (most common) drying medium.

Chapter 7 concentrates on energy conservation in textile finishing. In textile finishing it is essential to reduce energy consumption and thermal pollution. This leads to an essential study on the ways of limiting energy wastage and thermal pollution in textile finishing processes. Chapter 8 focuses on energy conservation in man-made fibre industry. Man-made fibres are spun and woven into a huge number of consumer and industrial products, including garments such as shirts and scarves. Various energy-efficiency technologies and measures in man-made fibre production are discussed in detail. Chapter 9 is devoted to recovery of reuse of chemicals in textile industry. Various methods for recovery of chemicals, PVA, caustic soda, solvents and chromium are discussed.

Chapter 10 concentrates on recycling and conservation of water. Water conservation and reuse can have tremendous benefits through decreased costs of purchased water and reduces costs for treatment of wastewater. Chapter 11 focuses on cogeneration. Cogeneration is an energy production process involving simultaneous generation of thermal (e.g., process steam) and electric energy by using single primary heating source.

Chapter 12 is devoted to energy efficient boilers. Various energy efficiency opportunities in boiler system can be related to combustion, heat transfer, avoidable losses, high auxiliary power consumption, water quality and blow down. Chapter 13 deals with efficient steam generation. Well the function of the steam distribution system is to get the steam to where it is needed and return the condensate to the boiler, doing both as efficiently as possible.

Chapter 14 focuses on waste heat recovery in textile industry. By implementing the waste heat recovery methods we can conserve the energy in the textile industries. The improvements in the boiler blow down, condensate recovery, feed water management and waste water recovery will minimise the energy losses and improve the performance of the thermal systems in textile industries. Like many other industrial units, textile mills are highly dependent upon electrical and heat energy. Some reactions which are necessary for chemical processing in cloth will not take place unless the components are heated. Heat is also needed for drying in textile mills. Keeping this in mind, chapter 15 concentrates on thermal and electrical conservation in textile industry. Chapter 16 focuses on energy efficient motors, gears, fans and compressors. When considering energy-efficiency improvements to motor systems, a systems approach incorporating pumps, compressors, and fans must be used in order to attain optimal savings and performance.

Chapter 17 is devoted to energy efficient pumps and V-belts. Pumping systems account for a significant percentage of energy consumption of the total industrial energy usage. Various energy saving options in pumping systems are discussed. Chapter 18 deals with energy efficient fuel oils and lubricants in textile industry. This chapter explains the various means and developments available to conserve fuel oil in textile plant particularly as applicable for typical fired boilers. Chapter 19 is devoted to energy saving in cooling towers. Energy efficiency can be obtained by using variable frequency drives by installing new nozzles to obtain a more uniform water pattern.

Chapter 20 concentrates on carbon footprint in textile industry. Carbon foot print can be described as the extent of damage caused to the environment due to some actions. It is the measure of severity of our activities on the environment, especially on the climate change. Chapter 21 focuses on role of nanotechnology in energy conservation. Nanotechnology provide the potential to enhance energy efficiency across all branches of industry and to economically leverage renewable energy production through new technological solutions and optimised production technologies.

Chapter 22 is devoted to energy audit in textile industry. Energy audit is the key to a systematic approach for decision making in the area of energy management. It attempts to balance the total energy inputs with its use and serves to identify all the energy streams in a facility.

I am thankful to Mr Sarvesh Devraj (Research associate – TERI, New Delhi) who helped me in editing the book. Appreciations are also extended to Mr Harinder Singh, Senior DTP operator, who drew and labelled the flow diagrams and worked long hours to bring the book on time. I am also thankful to the editorial team of Woodhead Publishing India Pvt. for their wholehearted cooperation in bringing out the book in time.

It may not be wrong to hold that this book on *Energy conservation in textile industry* is essential reading for professionals and students pursuing B. Tech, M. Tech in textile engineering and allied fields. Besides students, this book will prove useful to industrialists and consultants in the respective fields.

It has been prepared with meticulous care, aiming at making the book error-free. Constructive suggestions are always welcome from users of this book.

S C Bhatia
Prof. Puneet Mangla

1

Textile industry: An overview

1.1   Introduction

The textile industry is not a single entity but encompasses a range of industrial units which use a variety of natural and synthetic fibres to produce various fabrics. All over the world cotton has retained its dominant position not only because of its easy domestic availability but also due to the climatic conditions in the country which dictate the need for cotton based wear. The growth of the textile industry in any region is greatly influenced by the climatic conditions and availability of raw materials.

The textile industries are very complex in nature as far as varieties of products, process and raw materials are concerned. During production, the cloth has to pass through various processes and chemical operations like sizing, desizing, scouring, mercerising, bleaching, dying, printing, and finishing. In a textile industry, a number of dyes chemicals and auxiliary chemicals are used to impart desired quality in the fabrics. The wastewater of the industry is highly alkaline in nature and contains high concentration or BOD, COD, TDS and alkalinity. It can cause environmental problems unless it is properly treated before disposal. The industry also generates air pollution. Processing of fibres prior to and during spinning and weaving generates dust, lint, etc., which degrades working environment in the industry. Dust may cause respiratory diseases in workers. A chronic lung disease, byssinosis is commonly observed among workers exposed to cotton, flax and hemp dust. Besides this, there are a number of process operations including spinning weaving that produce noise to the tune of 90 dB (A).

1.2   Textile manufacturing processes

In general, the entire textiles manufacturing process can be described in five different stages (Fig. 1.1) of production, as given below:

1. Preparation of fibre natural (e.g., wool, cotton) and manmade, i.e., cellulosic (e.g., rayon, acetate) and synthetic (e.g., polyester, nylon).
2. Conversion of fibre into yarn (spinning).
3. Manufacturing of textile from yarn (weaving and knitting).
4. Colouring and finishing of textiles.
5. Garmenting by cutting and stitching.

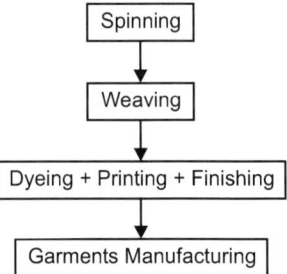

Figure 1.1: Process flow diagram of textile manufacturing.

The following sections describe each of these stages of production.

1.3   Preparation of fibre

1.3.1   Natural fibres

The natural fibres must be opened, blended, carded and/or combed and drafted before spinning. The main steps used for processing wool and cotton are briefed below. Although the equipment used for cotton is designed somewhat differently from that used for wool, the machinery operates in essentially the same fashion.

1.3.2   Man-made fibres

The manmade fibres (both synthetic and cellulosic) are manufactured by processes that simulate or resemble the manufacturing of silk (i.e., forcing a liquid through a small opening where the liquid solidifies to form a continuous filament). The main methods of fibre manufacturing are: (i) melt spining and (ii) dry-spinning and (iii) wet spinning. After the spinning process, the filaments are drawn to increase the orientation of the macromolecules and thereby the tensile strength of the yarns.

1.4   Spinning - conversion of fibre into yarn

The formation of spun yarn is done in spinning mills. Before spinning preparatory processes take place, the tasks of the processes are opening of the fibre bales, mixing of the fibres, cleaning, arrangement, paralleling of the fibres, drafting, and twining of the fibres to a yarn. Ring spinning is the most important technology (80% of worldwide yarn production). The open end technique is mostly used non-conventional spinning technology (Fig. 1.2).

The natural fibres as well as the man-made staple fibres are produced into yarns with different types of spinning systems. The kind of system used depends on the fibre length, fibre thickness and the end use of the product.

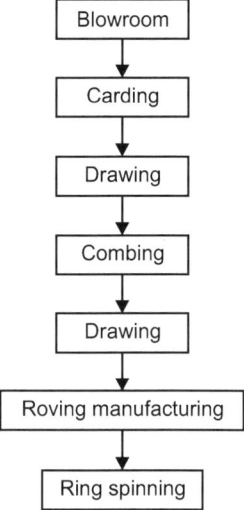

Figure 1.2: Flow diagram of spinning.

In all technologies mentioned below, the last step is carried out on ring spinning or non-conventional spinning machines:

1. Cotton spinning technologies [all fibre types (especially cotton) up to 40 mm length].
2. Worsted spinning (wool and long staple man-made fibres (especially polyester and polyacrylonitrile).
3. Semi worsted spinning [important for coarse wool and long staple man-made fibres (especially for polyamide and polyacrylonitrile].
4. Woolen spinning (universal technique for wool and fine man-made fibres).

The differences between these techniques are based on type and number of the spinning preparatory steps (drafting, combing, etc.).

Some yarn qualities are twisted (two or more yarns are twined up). From the environmental point of view, it is to be taken into account that during spinning and twisting lubricants and twisting oils may be applied, which are responsible for pollution loads in wastewater and off-gas in finishing (especially in pre-treatment processes).

1.4.1 Weaving and knitting-manufacturing of textile from yarn

Weaving

'Weaving' means to interlace two or more yarn systems crosswise and perpendicular. On the weaving machine (loom), the weft yarn is inserted into the lengthwise oriented warp yarns (shed). Before the weaving process starts, some preparatory processes have to be carried out. At first, the loom beam has to be prepared (Fig. 1.3).

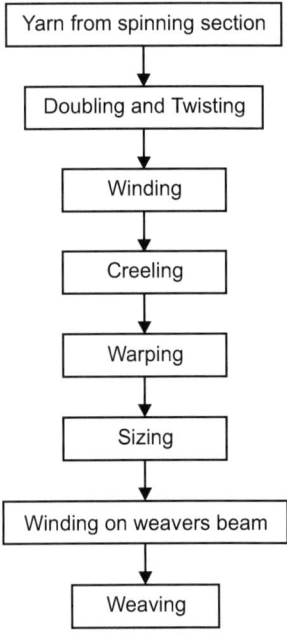

Figure 1.3: Flow diagram of weaving.

The warp yarns have to be assembled with the help of direct warping machines or sectional warping machines. Sectional warping is used for small highly patterned qualities. With respect to ecology it is important that warping oils are sometimes used in sectional warping and that, in most cases, beam warping is related to the sizing process. Most of the spun yarns and the main part of filament yarns have to be sized before weaving. Sizing is carried out in the weaving mill to protect the warp yarn during the weaving process from damage or break. The size forms a protective film on the warp yarn; protruding fibre ends causing loom stops are minimised. Sizing is done with help of sizing machines (slashers).

The yarns unreeled from warp beams are impregnated in the sizing box with the hot sizing liquor, surplus of size is removed by squeezing rollers, the yarns are subsequently dried and assembled to the loom beam.

In finishing the sizes (and also warping oils) have to be removed from the fabric leading to the main charge in the wastewater drainage of finishing mills. Due to different machinery manufacturers and different fabric qualities (fineness of yarns, fabric density, fabric pattern, etc.), different kinds of looms are used in weaving mills:

1. Eccentric looms (simple weave patterns).
2. Dobby machines (more kind of weave patterns).
3. Jacquard machines (most kind of weave patterns).

The weft insertion is carried out with the following techniques:

1. Shuttle.
2. Projectile.
3. Rapier.
4. Water-jet.
5. Air-jet.
6. Special weft insertion techniques.
7. Circles weave technique.

The size add-on on the warp yarns depends, besides some parameters of the yarn, on the type of weaving machine used, respectively, on the weft insertion rate. The woven textiles are used in all textile sectors (apparel, home textiles, and technical textiles).

Knitting

Knitted textiles are fabrics, which are made of yarns or yarn systems by stitch formation. Flat knitting, circular knitting, and warp knitting technologies exist. Besides the use in apparels (especially jumpers, underwear, hoses) and home textiles (especially net curtains), knitted textiles are also used for industrial textiles. Knitting oils used in the process are of ecological interest in downstream processing steps (especially pre-treatment in textile finishing mills).

1.4.2   Colouring and finishing of textiles

The processes of colouring and finishing are generally known as dyeing and printing process. The textile finishing mills are known as dyeing and printing mills. The main processes in textile dyeing and printing mills are summarised below. Depending on the demanded end-use properties of the textile all or only some of the above-mentioned processes are carried out.

Pre-treatment

In pre-treatment steps natural impurities on the textile raw material (greige, grey goods), e.g., by-products on cotton as waxes, proteins, etc., vegetable impurities on wool but also by-products from upstream production steps (preparation agents; sizing agents, etc.), and fibre specific by-products from man-made fibres (monomers, fibre solvents) are removed. These by-products together with the auxiliaries and chemicals used in pre-treatment cause a considerable ecological load in the wastewater as well as in the off-gas.

Dyeing

In dyeing, textiles are brought into contact with aqueous dyestuff solutions, variety of chemicals (salts, acids, etc.), and dyeing auxiliaries (surfactants, dispersing agents, levelling agents, etc.). The type and quantity of dyes, chemicals and auxiliaries are substrate specific and depend on the product quality (e.g., fastness properties), as well as on the type of installed machinery (Fig. 1.4). Colouration with dyes is based on physico-chemical equilibrium processes, namely diffusion and sorption of dye molecules or ions. These processes may be followed by chemical reactions in the fibres (e.g., reactive dyestuffs react with the fibres, metal complex dyestuffs generate complexes with the fibre molecules, vat and sulphur dyes have to be reoxidised). Dyeing is carried out in continuous and semi-continuous processes or batchwise.

Exhaust dyeing: In exhaust dyeing, the material is brought into contact with the dyeing liquor (water with dissolved or dispersed dyes and textile auxiliaries) in a dyeing machine. The dyes wear out from the dye bath and absorb on the fibres. The dyeing equilibrium depends on temperature, time, pH and textile auxiliaries. After dyeing, the exhausted dye bath is discharged and, depending on the kind of substrate, quality to be achieved, and dyestuff used, rinsing, soaping, and special after treatment processes take place. The dyeing of fabrics is possible in rope form (skein dyeing) or in full width. Different kinds of dyeing machines are available.

Important parameters in exhaust dyeing are:
1. Liquor ratio (kg textile to be dyed/L water used in dyeing bath).
2. Dyeing method (temperature/time curves; two bathes or one bath method in case of fibre mixtures).
3. Dyestuff type, auxiliaries.
4. Exhaustion degree of dyestuffs.
5. Amount of rinsing bathes and kind of after treatment needed.
6. Energy and cooling water consumption.

Semi-continuous dyeing: In semi-continuous dyeing (pad-jig, pad-batch, pad-roll), the fabric is impregnated in a padding machine with the dye-liquor and afterwards treated batch wise in a jigger or stored with slow rotation for several hours (pad-batch: at room temperature; pad-roll: at elevated temperature in a heating chamber) for fixation of the dyes on the fibre. After fixation, the material is washed and rinsed in full width on continuous washing machines.

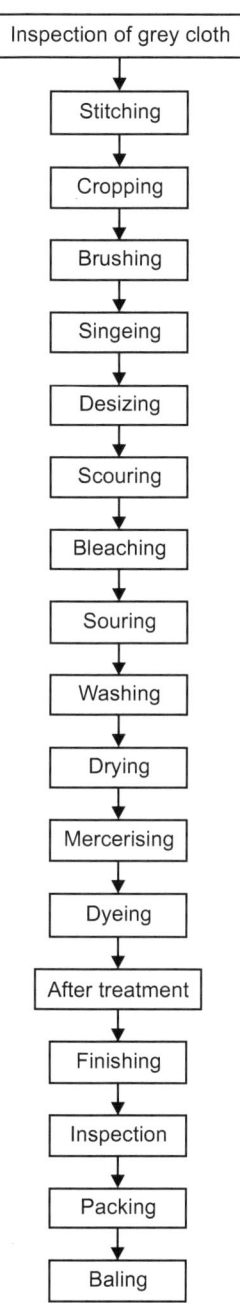

Figure 1.4: Flow diagram of dyeing.

Continuous dyeing: In continuous processes, the dyestuffs are applied in a padding mangle to the material with direct subsequent dye fixation by means of chemicals, heat, or steam followed by washing steps. Pad-steam processes (padding and fixation by steaming) and thermosol processes (padding of disperse dyes with subsequent heating) are commonly used.

1.4.3   Printing

Besides dyeing, colourisation in textile industry is possible by means of printing technologies, mainly used for multicolour patterns.

The most common printing technologies are:

1. Direct printing.
2. Discharge printing.
3. Resist printing.

Direct printing is the most common approach for applying a colour pattern (Fig. 1.5). It is done on white or previously dyed fabrics (generally in light colours to make the print stand out); in this case it is called overprinting. In discharge printing, a local destruction of a dye applied in a previously step takes place. If the etched areas become white, the process is called white discharge. If the printing paste contains reduction resistant dyes, the etched areas become coloured (coloured discharge technique). In the case of resist printing, a special printing paste (resist) is printed onto the textile to prevent dyestuff fixation. In subsequent dyeing, only the non-reserved areas are coloured.

Various printing paste application methods are applied, such as the following:

1. Roller printing.
2. Flat screen printing.
3. Rotary screen printing.
4. Transfer printing.
5. Ink jet (emerging technique).

Roller printing is a technique with recessed (engraved) printing forms. In flat screen-printing, the printing paste is transferred to the fabric through openings in specially designed screens. The openings correspond to the pattern when the printing paste is forced through by means of a squeegee.

Rotary screen printing uses the same principle, but instead of flat screens the printing paste is transferred through lightweight metal foil screens which are made in the form of cylinder rolls.

In transfer printing (mainly done on polyester), the environmental loads during textile printing are minimised. The patterns in transfer printing is

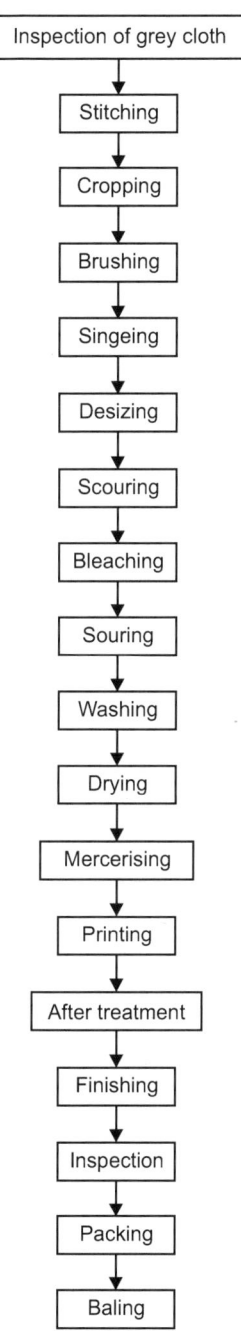

Figure 1.5: Flow diagram of printing.

transferred from a paper support to the fabric by means of heat. Ink jet printing on textiles can be carried out as jet printing on papers. Nowadays, this technique is used for small lots and patterning.

1.4.4   Finishing

Mechanical, thermal, and chemical treatments performed on fibres, yarns and fabrics after pre-treatment, dyeing or printing are summarised as 'finishing'. Finishing improves the functionality and the handle of the textile. Some finishing processes are specific for a special substrate (easy-care finishing on cotton, antistatic finishing for textiles made of man-made fibres).

Energy conservation in textile industry

2.1 Introduction

'Energy conservation' and 'energy efficiency' are often used interchangeably, but there are few differences. At the most basic level, energy conservation means using less energy and is usually a behavioural change, like turning lights off or setting thermostat lower. Energy efficiency, however, means using energy more effectively and is often a technological change. Energy efficiency measures the difference between how much energy is used to provide the same level of comfort, performance or convenience by the same type of product, building or vehicle.

Conservation certainly reduces energy use, but it's not always the best solution because it may impact comfort or safety as well. Efficiency, on the other hand, maintains the same level of output (e.g., light level, temperature) but uses less energy to achieve it. A combination of both energy conservation and energy efficiency measures yields an ideal solution.

2.2 Industrial sector energy efficiency

The industrial sector represents more than one third of both global primary energy use and energy-related carbon dioxide emissions. In developing countries, the portion of the energy supply consumed by the industrial sector is frequently in excess of 50% and can create tension between economic development goals and a constrained energy supply. Further, countries with an emerging and rapidly expanding industrial infrastructure have a particular opportunity to increase their competitiveness by applying energy-efficient best practices from the outset in new industrial facilities.

Integrating energy efficiency into the initial design or substantial redesign is generally less expensive and allows for better overall results than retrofitting existing industrial facilities, as is typically required in more developed countries. Conversely, failure to integrate energy efficiency in new industrial facility design in developing countries represents a large and permanent loss in climate change mitigation potential that will persist for decades until these facilities are scheduled for major renovation.

Despite the potential, policymakers frequently overlook the opportunities presented by industrial energy efficiency to have a significant impact on climate change mitigation, security of energy supply and sustainability. The common perception holds that energy efficiency of the industrial sector is too complex to be addressed through public policy and, further, that industrial facilities will achieve energy efficiency through the competitive pressures of the marketplace alone. Neither premise is supported by the evidence from countries that have implemented industrial energy efficiency programmes. The opportunities for improving the efficiency of industrial facilities are substantial, on the order of 20–30% even in markets with mature industries that are relatively open to competition. The principal business of an industrial facility is production, not energy efficiency. This is the underlying reason why market forces alone will not achieve industrial energy efficiency on a global basis, 'price signals' notwithstanding. High energy prices or constrained energy supply will motivate industrial facilities to try to secure the amount of energy required for operations at the lowest possible price. But price alone will not build awareness within the corporate culture of the industrial firm of the potential for the energy savings, maintenance savings and production benefits that can be realised from the systematic pursuit of industrial energy efficiency. It is this lack of awareness and the corresponding failure to manage energy use with the same attention that is routinely afforded production quality, waste reduction and labour costs that is at the root of the opportunity.

Industrial energy efficiency is dependent on operational practices, which change in response to variations in production volumes and product types. Due to this dependence, industrial energy efficiency cannot be fully realised through policies and programmes that focus solely on equipment components or specific technologies. Companies that actively manage their energy use seek out opportunities to upgrade the efficiency of equipment and processes because they have an organisational context that supports doing this wherever cost effective, while companies without energy management policies do not. Providing technology-based financial incentives in the absence of energy management will not result in significant market shifts because there is no organisational context to respond to and integrate the opportunity into ongoing business practice. A portfolio of industrial policies is needed that is designed to assist companies in developing this supporting context, while also providing consistency, transparency, engagement of industry in programme design and implementation, and, most importantly, allowance for flexibility of industry response. When these criteria are met, industry has shown that it can exceed expectations as a source of reductions in energy use and corresponding greenhouse gas (GHG) emissions, while continuing to prosper and grow.

Thus, industrial energy efficiency—or conversely, energy intensity, which is defined as the amount of energy used to produce one unit of a commodity—is determined by the type of processes used to produce the commodity, the vintage of the equipment used and the efficiency of production, including operating conditions. Energy intensity varies between products, industrial facilities and countries depending upon these factors.

Within industry, systems that support industrial processes that can be found to varying degrees in virtually all industrial sectors, regardless of their energy intensity. These industrial systems, which include compressed air, pumping and fan systems (referred to collectively as motor systems), steam systems and process heating systems are integral to the operation of industrial facilities, providing essential conversion of energy into energised fluids or heat required for production processes.

Motor and steam systems account for 15% and 38%, respectively, of global final manufacturing energy use, or approximately 46 EJ/year.

Because these systems typically support industrial processes, they are engineered for reliability rather than energy efficiency. Industrial systems that are oversized in an effort to create greater reliability, a common practice, can result in energy lost to excessive equipment cycling, less efficient part load operation and system throttling to manage excessive flow. Waste heat and premature equipment failure from excessive cycling and vibration are side effects of this approach that contribute to diminished, not enhanced, reliability.

More sophisticated strategies, made possible through the emergence of modern controls, create reliability through flexibility of response—and redundancy in the case of equipment failure—rather than by brute force. The energy savings can be substantial, with savings of 20% or more common for motor systems and 10% or more for steam and process heating systems.

2.2.1   Opportunities for industrial energy efficiency

Opportunities to improve industrial energy efficiency are found throughout the industrial sector. Assessments of cost-effective efficiency improvement opportunities in energy-intensive industries in the United States, such as steel, cement and paper manufacturing, found cost-effective savings of 16% to 18% even greater savings can often be realised in developing countries where old, inefficient technologies have continued to be used to meet growing material demands.

2.2.2   Barriers to industrial energy efficiency improvement

The decision-making process regarding investments in energy-efficient technologies is shaped by firm rules, corporate culture and the company's perception of its level of energy efficiency. Researchers found that most firms view themselves as energy efficient even when profitable improvements are available. Lack of knowledge or the limited ability of industrial commodity producers to research and evaluate information on energy-efficient technologies and practices is another barrier. Uncertainties related to energy prices or capital availability can lead to the use of stringent investment criteria and high hurdle rates for energy efficiency investments that are higher than the cost of capital to the firm. Capital rationing is often used within firms as an allocation means for investments, especially for small investments such as many energy efficiency retrofits. The relatively slow rate which industrial capital stock turns over can prove to be a barrier to adoption of energy efficiency improvements since new stock is typically more energy-efficient than existing facilities. Another barrier is the perceived risk involved with adopting new technology since reliability and maintenance of product quality are extremely important to commodity producers. Optimising industrial systems for energy efficiency is not taught to engineers and designers at university—it is learned through experience. Systems are designed to maintain reliability at the lowest first cost investment, despite the fact that operating costs are often 80% or more of the life cycle cost of the equipment. Facility plant engineers are typically evaluated on their ability to avoid disruptions and constraints in production processes, not energy-efficient operation. Equipment suppliers also have little incentive to promote more energy-efficient system operation, since commissions increase when equipment size is scaled upward and educating a customer to choose a more efficient approach requires extra time and skill.

Plant engineering and operations staff frequently experience difficulty in achieving management support. Industrial managers are rarely drawn from the ranks of facilities operation—they come from production and often have little understanding of supporting industrial systems. This situation is further exacerbated by the existence of a budgetary disconnect in industrial facility management between capital projects (including equipment purchases) and operating expenses. In addition, most optimised industrial systems lose their initial efficiency gains over time due to personnel and production changes. Detailed operating instructions are not integrated with quality control and production management systems. Without well documented maintenance procedures, the energy efficiency advantages of high efficiency components can be negated by clogged filters, failed traps and malfunctioning valves.

2.2.3 Energy efficiency

Typically, the process for setting energy efficiency or GHG emission reduction targets requires a preliminary assessment of the energy efficiency or GHG mitigation potential of each industrial facility, which includes an inventory of economically viable measures that could be implemented.

2.3 Causes of the energy crisis

The energy crisis is the concern that the world's demands on the limited natural resources that are used to power industrial society are diminishing as the demand rises. These natural resources are in limited supply. Governments and concerned individuals are working to make the use of renewable resources a priority and to lessen the irresponsible use of natural supplies through increased conservation.

The energy crisis is a broad and complex topic. Most people don't feel connected to its reality unless the price of gas at the pump goes up or there are lines at the gas station. The energy crisis is something that is ongoing and getting worse, despite many efforts. The reason for this is that there is not a broad understanding of the complex causes and solutions for the energy crisis that will allow for an effort to happen that will resolve it.

'An energy crisis is any great bottleneck (or price rise) in the supply of energy resources to an economy. In popular literature though, it often refers to one of the energy sources used at a certain time and place, particularly those that supply national electricity grids or serve as fuel for vehicles.'

2.3.1   Factors related to energy crisis

Some of the factors related to energy crisis are discussed below:

Over consumption: The energy crisis is a result of many different strains on our natural resources, not just one. There is a strain on fossil fuels such as oil, gas and coal due to over consumption – which then in turn can put a strain on our water and oxygen resources by causing pollution.

Over population: Another cause of the crisis has been the steady increase in the world's population and its demands for fuel and products. No matter what type of food or products you choose to use – from fair trade and organic to those made from petroleum products in a sweet shop – not one of them is made or transported without a significant drain on our energy resources.

Poor infrastructure: Ageing infrastructure of power generating equipment is yet another reason for energy shortage. Most of the energy producing firms keep on using outdated equipment that restricts the production of energy. It is the responsibility of utilities to keep on upgrading the infrastructure and set a high standard of performance.

Unexplored renewable energy options: Renewable energy still remains unused in most of the countries. Most of the energy comes from non-renewable sources like coal. It still remains the top choice to produce energy. Unless we give renewable energy a serious thought, the problem of energy crisis cannot be solved. Renewable energy sources can reduce our dependance on fossil fuels and also helps to reduce greenhouse gas emissions.

Delay in commissioning of power plants: In few countries, there is a significant delay in commissioning of new power plants that can fill the gap between demand and supply of energy. The result is that old plants come under huge stress to meet the daily demand for power. When supply doesn't matches demand, it results in load shedding and breakdown.

Wastage of energy: In most parts of the world, people do not realise the importance of conserving energy. It is only limited to books, internet, newspaper ads, lip service and seminars. Unless we give it a serious thought, things are not going to change anytime sooner. Simple things like switching off fans and lights when not in use, using maximum daylight, walking instead of driving for short distances, using CFL instead of traditional bulbs, proper insulation for leakage of energy can go a long way in saving energy.

Poor distribution system: Frequent tripping and breakdown are result of a poor distribution system.

Major accidents and natural calamities: Major accidents like pipeline burst and natural calamities like eruption of volcanoes, floods, earthquakes can also cause interruptions to energy supplies. The huge gap between supply and demand of energy can raise the price of essential items which can give rise to inflation.

Wars and attacks: Wars between countries can also hamper supply of energy specially if it happens in Middle East countries like Saudi Arabia, Iraq, Iran, Kuwait, U.A.E. or Qatar. That's what happened during 1990 Gulf war when price of oil reached its peak causing global shortages and created major problem for energy consumers.

Miscellaneous factors: Tax hikes, strikes, military coup, political events, severe hot summers or cold winters can cause sudden increase in demand of energy and can choke supply. A strike by unions in an oil producing firm can definitely cause an energy crisis.

2.3.2 Possible solutions of the energy crisis

Move towards renewable resources: The best possible solution is to reduce the world's dependence on non-renewable resources and to improve overall conservation efforts. Much of the industrial age was created using fossil fuels, but there is also known technology that uses other types of renewable energies such as– steam, solar and wind. The major concern is not so much that we will run out of gas or oil, but that the use of coal is going to continue to pollute the atmosphere and destroy other natural resources in the process of mining the coal that it has to be replaced as an energy source. This is not easy as many of the leading industries use coal, not gas or oil, as their primary source of power for manufacturing.

1. Buy energy efficient products.
2. Lighting controls.

3. Easier grid access.
4. Energy simulation.
5. Perform energy audit.
6. Common stand on climate change.

2.4   Need of energy conservation

The amount of energy consumption in the entire world has been increased, accompanied by economic development of each country. Many energy resources used throughout the world today are still fossil fuels such as oil, coal and natural gas. If energy consumption continues to increase at the same rate as today, exhaustion of resources would occur in the near future.

Additionally, as a result of mass consumption of fossil fuels, global warming caused by an increasing amount of CO_2 emissions in the air has been occurring at rapid speed, which is one of the most crucial global issues.

As effective counter measures against global issues such as future exhaustion of resources and global warming, the necessity for the promotion of energy conservation in the international level has been increasingly emphasised. In recent years, various policies of energy conservation have been implemented in many countries. Also, the international framework based on the Kyoto Protocol was established and it has promoted activities towards tackling global warming along with the ratification by Russia.

At the same time, many countries still have a strong tendency to focus on economic development rather than environmental measures. It cannot be said that energy conservation is the issue that society wants first and most. This tendency is seen in some developed countries such as the U.S., but especially in developing countries, policy priority is not given to energy conservation.

Energy has an important function. It is the central force behind our productivity, our leisure and our environment. There is a strong correlation between energy use per person and standard of living in each economy. A higher per capita energy consumption means a higher per capita gross national product. Energy is an indispensable component of industrial product, employment, economic growth, environment and comfort. Low cost energy was abundant in the past. Energy cost was only a very small fraction of the cost of finished product. Use of low cost energy for home comfort became very predominant. The subsequent increase in oil prices increased the energy cost in every sector, domestic, commercial, industrial, etc. The per capita energy consumption in India is very low as compared to that in advanced countries. However our energy resources are fast getting depleted. Thus energy saving or conservation is essential in developed as well as developing countries.

Meaning and principles of energy conservation: Energy conservation means using energy more efficiently or reducing wastage of energy. It is important that any energy conservation plan should only to try to eliminate wastage of energy without in any way affecting productivity and growth rate. A small decrease in convenience or comfort can be tolerated. Energy conservation usually requires new investment in more efficient equipment to replace old inefficient ones. Thus energy conservation can result in more job opportunities, lower costs, cheaper and better products, etc.

There are two principles of energy conservation planning which are discussed below:

1. Maximum energy efficient: A device, system or process is working at maximum efficiency when maximum work is done for a given magnitude of energy input. Only a part of the input energy is converted into useful work. The remainder is lost in energy conversion and transfer process and energy discharge.

   Work = Energy input – Energy loss in energy conversion transfer process and energy discharge.

2. Maximum cost effectiveness in energy use: Implementation of energy conservation entails additional investment. This investment increases as more and more energy conservation measures are adopted. Because of implementation of these measures the fuel costs decreases as extent of conservation is increased. The total cost per unit output is the sum of annual charges on investment per unit output and fuel costs per unit output. Evidently maximum cost effectiveness in energy use is obtained when total costs are the least.

2.5   Energy conservation in textile industry

The conservation of energy is an essential step we can all take towards over coming the mounting problems of the worldwide energy crisis and environmental degradation. In the textile industry, appreciable amounts of energy could be saved or conserved by regulating the temperature in the steam pipes, adjusting the air/fuel ratio in the boilers and installing heat exchangers using warm waste water.

The textile industry is one of the most complicated manufacturing industries because it is a fragmented and heterogeneous sector dominated by small and medium enterprises (SMEs). Characterising the textile manufacturing industry is complex because of the wide variety of substrates, processes, machinery and components used and finishing steps undertaken. Different types of fibres or yarns, methods of fabric production and finishing processes (preparation, printing, dyeing, chemical/mechanical finishing and coating), all inter-relate in producing a finished fabric. Energy is one of the main cost factors in the textile industry. Especially in times of high energy price volatility, improving energy-efficiency should be a primary concern for textile plants. There are various energy-efficiency opportunities that exist in every textile plant, many of which are cost-effective.

However, even cost-effective options are not often implemented in textile plants mostly because of limited information on how to implement such energy-efficiency measures, especially given the fact that a majority of textile plants are categorised as SMEs and hence they have limited resources to acquire this information. Know-how on energy-efficiency technologies and practices should, therefore, be prepared and disseminated to textile plants.

The textile industry uses large quantities of both electricity and fuels. The share of electricity and fuels within the total final energy use of any one country's textile sector depends on the structure of the textile industry in that country. For example, in spun yarn spinning, electricity is the dominant energy source, whereas in wet-processing the major energy source is fuels.

Energy use in the spinning process: Electricity is the major type of energy used in spinning plants, especially in cotton spinning systems. If the spinning plant just produces raw yarn in a cotton spinning system and does not dye or fix the produced yarn, the fuel may just be used to provide steam for the humidification system in the cold seasons for pre-heating the fibres before spinning them together. Therefore, the fuel used by a cotton spinning plant highly depends on the geographical location and climate in the area where the plant is located.

Energy use in wet-processing: Wet-processing is the major energy consumer in the textile industry because it uses a high amount of thermal energy in the forms of both steam and heat. The energy used in wet-processing depends on various factors such as the form of the product being processed (fibre, yarn, fabric, cloth), the machine type, the specific process type, the state of the final product, etc.

The spinning consumes the greatest share of electricity (41%) followed by weaving (18%). Wet-processing preparation (desizing, bleaching, etc.), and finishing together consume the greatest share of thermal energy (35%). A significant amount of thermal energy is also lost during steam generation and distribution (35%). These percentages will vary by plant.

Energy-efficiency improvement opportunities in the textile industry: This analysis of energy-efficiency improvement opportunities in the textile industry includes both opportunities for retrofit/process optimisation as well as the complete replacement of the current machinery with state-of-the-art new technology. However, special attention is paid to retrofit measures since state-of-the-art new technologies have high upfront capital costs and therefore the energy savings which result from the replacement of current equipment with new equipment alone in many cases may not justify the cost.

However, if all the benefits received from the installation of the new technologies, such as water savings, material saving, reduced waste and waste water, reduced redoing, higher product quality, etc., are taken into account, the new technologies are more justifiable economically.

Energy conservation in spinning process

3.1   Introduction

Spinning is the first steps of textile product processing. The process of making yarns from the textile fibre is called spinning. Spinning is the twisting together of drawn out strands of fibres to form yarn, though it is colloquially used to describe the process of drawing out, inserting the twist and winding onto bobbins. There are different types of spinning, the most commonly forms of spinning are: ring, rotor spinning, air jet, friction, etc.

Spinning is a major part of the textile industry. It is part of the textile manufacturing process where three types of fibre are converted into yarn, then fabrics, which undergo finishing processes such as bleaching to become textiles. The textiles are then fabricated into clothes or other products. There are three industrial processes available to spin yarn and a handicraft community who use hand spinning techniques. Spinning is the twisting together of drawn out strands of fibres to form yarn, though it is colloquially used to describe the process of drawing out, inserting the twist and winding onto bobbins.

In simple words, spinning is a process in which we convert fibres by passing through certain processes like blow room, carding, drawing, combing, simplex, ring frame and finally winding into yarns. These yarns are then wound onto the cones.

Artificial fibres are made by extruding a polymer through a spinneret into a medium where it hardens. Wet spinning (rayon) uses a coagulating medium. In dry spinning (acetate and triacetate), the polymer is contained in a solvent that evaporates in the heated exit chamber. In melt spinning (nylons and polyesters) the extruded polymer is cooled in gas or air and sets. All these fibres will be of great length, often kilometers long.

Natural fibres are either from animals (sheep, goat, rabbit, silk-worm), mineral (asbestos), or from plants (cotton, flax, sisal). These vegetable fibres can come from the seed (cotton), the stem (known as bast fibres: flax, hemp, jute) or the leaf (sisal). Without exception, many processes are needed before a clean even staple is obtained – each with a specific name. With the exception of silk, each of these fibres is short, being only centimetres in length and each has a rough surface that enables it to bond with similar staples.

Artificial fibres can be processed as long fibres or batched and cut so they can be processed like a natural fibre.

Ring-spinning is the most common spinning method in the world. Other systems include air-jet and open-end spinning. Open-end spinning is done using break or open-end spinning. This is a technique where the staple fibre is blown by air into a rotor and attaches to the tail of formed yarn that is continually being drawn out of the chamber.

3.2 Energy-efficiency technologies and measures in the spun yarn spinning process

Energy-efficiency technologies and measures in the spun yarn spinning process are given below.

3.2.1 Preparatory process

Installation of electronic roving end-break stop-motion detector instead of pneumatic system

In a simplex (roving) machine, the roving end-break system can be converted from a pneumatic suction tube detector to a photoelectric stop-motion system end-break detector in order to save energy. This measure is implemented in many textile plants around the world. The average energy saving reported from implementation in two Indian spinning plants is 3.2 MWh/year/machine with an average investment cost of about US$180 per roving machine.

High-speed carding machine

This machine is used in the secondary processing of raw cotton. The machine separates the lumps of small fibres that result from the disentangling of tufts in the 'opening-and-picking' stage of primary processing and simultaneously removes impurities, lint balls and short fibres, improving the arrangement of good quality fibres in the longitudinal direction and producing fibre bundles (slivers) in strands.

This new carding machine is large and each machine consumes considerable amounts of electricity. On the other hand, since productivity is high, 1/3 the number of new machines and half the total power can produce the same production capacity as ordinary carding machines.

For instance, twelve conventional machines requiring 27 kW/machine can be replaced by four of the new machines requiring 41 kW/machine and thus results in power-savings of 160 kW. There are many examples of installation of new carding machines in major plants throughout Japan and this technology is certainly applicable in any developing country. The capital cost of the new carding machine is about US$100,000. The payback period for the investment is about 1.3 years.

3.2.2   Ring frame

Use of energy-efficient spindle oil

The incorporation of a dispersant additive system to the mineral-based spindle oil may result in energy savings of up to 3% when compared to conventional oils. The amount of actual savings will depend upon the condition of the machinery and their operation condition. Energy saving also can be achieved by using light weight spindles. Synthetic-based spindle oils (energy-efficient grades) along with certain metal compatibility additives may result in higher energy savings, in the range of 5–7% depending upon viscosity. The energy-efficiency potential of particular oil can be assessed in two ways:

1. The reduction in electricity consumption.
2. The reduction in bolster temperature rise over ambient: energy saving oils result in lesser temperature increases.

While selecting any energy saving spindle oil, one should carefully evaluate important characteristics related to the service life of the oil, i.e., temperature rise, thermal stability, metal compatibility, sludge forming tendency and anti-wear/antifriction properties.

Optimum oil level in the spindle bolsters

The electricity consumption in the ring frame increases with an increase in the oil level in the bolsters because of resistance caused by the oil. Also, an excessively high oil level in the bolster may disturb the proper running of the spindle. Normally, 75% of the bolster capacity is filled with oil. The usual method of determining the depth of oil level in the bolster is by lifting the spindle and observing the oiliness of the spindle blade. The correct or exact amount of oil for each type of spindle insert could be assured by using a dipstick. The dipstick has two distinct markings, i.e., the bottom marking for minimum and the top marking for maximum oil levels.

Replacement of lighter spindle in place of conventional spindle in ring frame

Ring frames are the largest energy consumer in the ring spinning process. Within a ring frame, spindles rotation is the largest energy consumer. Thus, the weight of the spindles is directly related to the energy use of the machine. There are so-called high efficiency spindles on the market which are lighter than the conventional spindles and hence use less energy. A spinning plant in India replaced the conventional spindles with lighter weight ones in their ring frames and on average saved 23 MWh/year/ring frame. The investment cost of this measure was around US$13,500 per ring frame.

Synthetic sandwich tapes for ring frames

Synthetic sandwich spindle tapes are made of polyamide, cotton yarn and a special synthetic rubber mix. Sandwich tapes run stable, have good dimensional stability, don't break, result in less weak-twist yarn, do not cause fibre sticking and are made of soft and flexible tape bodies. Because of these special characteristics, these tapes offer 5–10% energy saving. Based on an assessment conducted for an Indian spinning plant, replacing cotton tapes with synthetic sandwich spindle tapes can result in average savings of 8 MWh/ring frame/year (92.5 kWh/T/year). The capital cost of replacement is US$540 for each ring frame (with a payback period of about 10 months).

In another spinning plant in India, the installation of energy-efficient tapes in ring frames in average resulted in energy savings of 4.4 MWh/year/ring frame for a capital cost of US$683 per ring frame.

The cause of the difference between energy saving of these two plants could be the difference between the size of the ring frames in each plant (number of spindles in each ring frame).

Optimisation of ring diameter with respect to yarn count in ring frames

Ring diameter significantly influences the energy use of the ring frame. Larger ring diameters facilitate higher bobbin content with a heavier package, resulting in excess energy consumption. A reduction of about 10% in bobbin content lowers ring frame energy intensity by about 10%. For finer yarn counts, 38 millimeter (mm)/36 mm diameter and for medium yarn counts, 40 mm ring diameters are recommended. The cost of implementation is about US$1600 for a long length ring frame of 1008 spindles and the payback period is about 2 years.

False ceiling in ring spinning section

The spinning process needs to be under a maintained temperature and humidity. This is done in a humidification area within plants. The energy used by the humidification facility is directly related to the volume of the facility where the spinning process is carried out. The use of a false ceiling can help to reduce this volume, thereby reducing energy consumption. In a spinning plant in India, the volume of a spinning hall with 15000 spindles reduced energy use through the installation of a false ceiling under the hall's roof. This measure resulted in 125 MWh/year electricity savings (8 kWh/spindle/year). The capital cost for this renovation was about US$11000 (US$0.7/spindle).

Installation of energy-efficient motor in ring frame

Electric motors installed in the ring frames have the highest possible efficiency. Even a slight efficiency improvement in ring frame motors could result in significant electricity savings that could pay back the initial investment in a short period.

Installation of energy-efficient excel fans in place of conventional aluminium fans in the suction of ring frame

Ring frames have suction fans, which are used to collect fibres when a yarn break occurs. Energy-efficient excel fans could be installed in place of conventional aluminium fans in the suction system of ring frames. The average electricity savings from the implementation of this measure is reported to be between 5.8 and 40 MWh/year/fan with a capital cost of about US$195–310 per fan.

Use of light weight bobbins in ring frame

In ring frames, yarn is collected on bobbins. Bobbins are rotated by spindles upon which they sit. The rotating of spindles is the highest energy consumption activity in ring machines. The heavier the bobbins are, the more energy is required for the rotation of bobbins and hence spindles. Nowadays, the use of lighter bobbins in place of conventional ones is getting more attention. In a spinning plant. Replacement of 30–35 gram bobbins (cops) with 28 gram bobbins resulted in average electricity savings of 10.8 MWh/year/ring frame (assuming 12 doff a day). The capital cost for this retrofit measure was US$660 per ring frame.

High-speed ring spinning frame

This machine has an increased operating speed by 10–20% with similar power consumption as compared to conventional equipment. As a result, the power requirement is 36.0–40.5 kW in comparison with that of 45 kW for conventional ring spinning machines for the same production capacity. Furthermore, this equipment adopts an energy saving spindle that uses a small diameter warp, which contributes a power savings of approximately 6%.

Installation of a soft starter on motor drive of ring frame

The starting current drawn by an induction motor is directly proportional to the applied voltage. A soft-starter is designed to make it possible to choose the lowest voltage possible (the 'pedestal voltage') at which the motor can be started–the lowest voltage being dependent on the load on the motor. The voltage is ramped up from this pedestal level to full voltage within a preset time. Pedestal voltage and ramp-up time can be set at the site. It is also possible to provide controlled-torque soft-starts with current limit options. The soft-starter is also suited to situations where a smooth start is desirable to avoid shocks to the drive system or where a gradual start is required to avoid damage to the product/process/drive system and accessories.

In spinning plants, a soft-starter can reduce the costs incurred by yarn breaks on a ring frame when its motor starts after each doff, as smooth starts and gradual acceleration of motors eliminate shocks during starting. Average electricity savings reported from the implementation of this measure on ring frames is about 1–5.2 MWh/year per ring frame. The payback period of this measure is about 2 years. In addition to the electricity savings, the other advantages of this measure are a reduction in the maximum power demand and an improvement in the power factor.

3.2.3   Windings, doubling and finishing process

Installation of variable frequency drive on autoconer machine

Autoconer is the name of the machine which usually is used subsequent to ring frames in the yarn spinning process. The small bobbins of yarn are rewound onto larger cones by this machine. The installation of variable frequency drives (VFD) on an Autoconer's main motor can help maintain a constant vacuum and save energy. The adoption of this measure in spinning plant resulted in electricity savings of 331.2 MWh/year. The investment cost associated with this measure was about US$19,500.

Intermittent mode of movement of empty bobbin conveyor in the autoconer/cone winding machines

The continuous movement of empty bobbin conveyor belts can be converted into an intermittent mode of movement. This measure results in not only substantial energy saving but also results in maintenance cost savings and waste reduction. In a spinning plant in India, they converted the continuous conveyor system to an intermittent mode whereby the belts are running for 6 minutes only and stopping for 54 minutes in an hour. This resulted in electricity savings of 49.4 MWh/year. The investment cost associated with this measure was about US$1100.

Modified outer pot in two-for-one (TFO) machines

The process of twisting and doubling is an indispensable means of improving certain yarn properties and satisfying textile requirements that cannot be fulfilled by single yarns. The method of twisting two or more single yarns is called doubling, folding or ply twisting. Such yarns are designated as doubled yarn, folded yarn or plied yarn and the machines which conduct this work are called doublers, ply-twisters or two-for-one (TFO) twisters. Traditionally, ring doublers were used for ply twisting spun yarns and uptwisters were used for twisting filament yarns. Nowadays, TFO twisters are gaining world-wide acceptance in both the spun yarn and filament yarn sectors mainly because of their inherent advantages like the production of long lengths of knot-free yarns, which facilitates better performance in the subsequent processes and results in higher productivity.

In two-for-one twisting machines, the balloon tension of yarn accounts for about 50% of total energy consumption. The balloon diameter can be reduced with a reduction in yarn tension by providing a modified outer pot. This measure saves about 4% of total energy consumption in TFOs. Research shows that there is no deterioration in yarn quality.

Optimisation of balloon setting in two-for-one (TFO) machines

It has been observed above that TFOs consume less electricity at lower balloon settings. Balloon size can be optimised by taking account of various studies with respect to different yarn counts. An Indian textile plant saved about 250 MWh/year by optimising the balloon setting of its TFO machines without any investment required the number of TFO machine in which this measure was applied is not reported.

Replacing the electrical heating system with steam heating system for the yarn polishing machine

After applying a liquid polishing material on yarns, the yarn becomes wet and needs drying. In some plants yarn polishing machines use electrical heaters. These electrical heaters can be replaced by steam heaters which can reduce overall energy use. Steam consumption can increase by about 31.7 T steam/year for each machine, while electricity use declined in average by about 19.5 MWh/year/machine. The investment cost for this retrofit measure is reported to be about US$980 for each machine (with a payback period of about half a year).

3.2.4   Air conditioning and humidification system

Replacement of nozzles with energy-efficient mist nozzles in yarn conditioning room

In some textile plants the yarn cones are put in a yarn conditioning room in which yarn is kept under a maintained temperature and humidity. In such rooms, usually water is sprayed in to the air to provide the required moisture for the yarn to improve its strength, the softness and quality of the yarn and to increase its weight. The type of nozzles used for spraying the water can effectively influence the electricity use of the yarn conditioning system. In a case-study in India, by replacing jet nozzles with energy-efficient mist nozzles in a yarn conditioning room, resulting in 31 MWh/year electricity savings.

Installation of variable frequency drive (VFD) for washer pump motor in humidification plant

In humidification plants, an inverter can be installed on washer pump motors with auto speed regulation, which can be adjusted to meet the required humidity levels. Usually the pumps runs at 100% speed and humidity is controlled by by-passes, resulting in wasted energy. With VFDs, pump motor speed can be adjusted according to the requirements of the humidification plant. This could result in electricity saving as high as 20 MWh/year with an investment cost of about US$1100.

Replacement of the existing aluminium alloy fan impellers with high efficiency fibreglass reinforced plastic (FRP) impellers in humidification fans and cooling tower fans of spinning mills

Axial fans are widely used for providing required airflow in sections of textile industry plants. These sections include cooling towers and air-conditioning, ventilation and humidification systems. Optimal aerodynamic design of FRP (fibreglass reinforced plastic) fan impellers provides higher efficiencies for any specific application. A reduction in the overall weight of fans also extends the life of mechanical drive systems. Fans with FRP impellers require lower drive motor ratings and light duty bearing systems. Fans with FRP impellers consume less electricity compared to fans with aluminium alloy impellers under the same working conditions. In a case-study, a spinning plant replaced the impellers of 17 fans in the humidification system and cooling towers with FRP impellers. This retrofit measure resulted in average savings of 55.5 MWh/year/fan. The average investment cost of the replacement of impellers for each fan was about US$650.

Installation of VFD on humidification system fan motors for the flow control

Temperature and humidity levels must be closely monitored and maintained for textile processes (especially spinning and weaving) so that yarns will run smoothly through the processing machines; a well functioning ventilation system is imperative to the plant's successful operation. Ventilation systems use supply fans (SFs) and return fans (RFs) to circulate high humidity air to maintain proper ambient conditions, cool process machinery and control suspended particulate and airborne fibres. Initially, the mixture of return air and fresh air is cleaned, cooled and humidified by four air washers. This air is then supplied to the facility by the SFs and distributed to the plant through ceiling mounted ducts and diffusers, producing required temperatures and relative humidity levels. The RFs then pull air through the processing machines into a network of underground tunnels that filter out suspended particles and fibres, usually through rotary drum filters on the inlet of each RF.

While the psychometric qualities and volumes of air supplied and returned from each area remain relatively constant in the system, seasonal variations occasionally cause minor changes in ventilation rates. In addition, different products result in changing heat loads in the plant due to a varying number of running motors and/or loads on the motor. Factors that influence the pressure, volume, or resistance of the system directly impact the fan energy requirements. Therefore, air density, changes to damper positions, system pressure and air filter pressure drops, supply and return air system interaction and parallel fan operation all affect how much energy the fans require and must be monitored to ensure the efficient functioning of the system. Variable inlet guide vanes (VIVs) and outlet dampers usually initially control the system's air flow and these are highly inefficient. Setting these devices is imprecise, resetting the openings can be done manually or automatically and the VIVs and dampers can experience corrosion problems due to the high humidity in the air.

VFDs can be installed on flow controls; these devices control fan speed instead of changing the dampers' position. Thus, damper control is no longer necessary, so in the use of VFDs fan control dampers are opened 100%, thereby save electricity use by the fans. The average electricity saving reported for this retrofit in a plant in the US is 105 MWh/year/fan with the cost of US$8660, whereas an Indian textile plant has reported the average energy saving of 18 MWh/year/fan with the cost of US$1900. The saving and cost of the measure depend on various factors such as the size of the fan, the operating conditions, the climate, the type of VFD used, etc.

Installation of VFD on humidification system pumps

Pumps of humidification systems in spinning plants are usually running at full strength and throttling valves are used for controlling relative humidity. In place of throttling valves, variable frequency drives can be installed for controlling relative humidity and thereby the speed of the pumps can be reduced. This retrofit measure resulted in electricity savings equal to 35 MWh/year in a worsted spinning plant.

Energy-efficient control system for humidification system

On average, the humidification plants in textile industry consume about 15% to 25% of the total energy of the plant. Energy-efficient control systems consists of variable speed drives for supply of air fans, exhaust air fans and pumps in addition to control actuators for fresh air, recirculation and exhaust dampers. Energy savings in the range of 25% to 60% is possible by incorporating such control systems in the plants depending on the outside climate. These measures can be easily retrofitted in the existing humidification plants (both in automatic and manually-operated humidification plants) and the entire system can be controlled through a central computer. The estimated energy saving by implementing this control system is about 50 MWh/year.

3.2.5 General measures for spinning plants

Energy conservation measures in overhead travelling cleaner (OHTC)

Textile plants, especially spinning and weaving plants, usually need to effectively manage the waste (fluff) generated during fibre processing, which affects the quality of the outgoing yarn/fabric. It is imperative for textile plants to have control over waste removal out of the processing area to ensure best yarn and fabric quality. Fluff removal and machine cleaning can be accomplished with the support of overhead travelling cleaners (OHTC), which use an ancillary drive associated with spinning and weaving sections of textile plants. A common waste collection system (WCS) is an independent sub-system designated to collect waste from groups of OHTC(s).

In modern mills, one overhead travelling cleaner serves every 1008/1200-spindle ring spinning frame. It moves on rails at a speed of about 16 meters per minute. It takes about 140 seconds to move from one end of the ring frame to the other end. OHTC is continuously blowing/sucking off air and waste in and around different component parts of the machine during its traverse motion. In general, one overhead travelling cleaner consumes about 17,000 kWh per year. Some technology providers have developed different innovative methods of energy savings in overhead cleaners.

Some of these methods are described below:

1. Timer-based control system for overhead travelling cleaners (OHTC): An energy-efficient control system using timer circuits can be introduced in addition to a main contactor provided in the control box to start and stop the OHTC whenever it touches the ends of the ring frame over which it moves in a linear path. An off timer can be incorporated, with a feature of extending the delay of operation for 0 to 30 min in a stepped manner. The case-study showed that the operating time can be reduced by about 33%; hence energy consumption can be reduced by 33% when compared with base conditions. This method can be adopted for plants processing fine counts in which dust liberation is less. This system is not suited for plants processing coarser counts, as fly liberation will be greater. The average electricity savings reported from a case-study were 5.8 MWh/year per OHTC. The investment cost was about US$180 per OHTC unit.

2. Optical control system for overhead travelling cleaners (OHTC): An optical sensor to sense the position of the OHTCs on the ring frames can also be used. This system will start running the blower fan of the WCS only during the required operation time. By using the OHTC the energy consumption can be reduced by 41% when compared with base conditions. This can be translated to average electricity savings equal to 5.3 MWh/ year per OHTC. The investment required for this retrofit measure was US$980 per unit.

Energy-efficient blower fans for overhead travelling cleaner (OHTC)

Existing blower fans of OHTCs can be replaced by energy-efficient fans with smaller diameters and less weight. An energy savings of about 20% is achievable with a quick payback period of less than 6 months. A case-study implementing this measure reported average electricity savings of 2 MWh/year for each fan. The investment cost of the retrofit was about US$100 per fan.

Improving the power factor of the plant (reduction of reactive power)

There are many electric motors in a spinning plant that can cause reactive power. Therefore, reducing reactive power by improving the power factor of the plant is an important measure in reducing energy use and costs. For example, an Indian spinning plant replaced low value capacitors and added new capacitors where required. In this process system losses were reduced. The reported electricity saving is 24.1 MWh/year. The investment cost is reported to be around US$3300.

Replacement of ordinary 'V–belts' by cogged 'V–belts'

In textile plants, many motors are connected to the rotating device with pulleys and belts. In many cases, a V-belt is used to transfer the motion. Ordinary V-belts can be replaced with cogged V-belts to reduce friction losses, thereby saving energy. The implementation of such modification on 20 V-belt drives in a spinning plant in India resulted in electricity savings equal to 30 MWh/year. The capital cost of this modification was US$244.

Energy conservation in weaving process

4.1   Introduction

Weaving and knitting are the two most common processes of making cloth. Of these two processes, weaving is the most common method, although new and improved knitting machines make cloth quickly, satisfactorily and with attractive patterns. The majority of the fabric production is based on the woven fabrics. Weaving is a method of textile production in which two distinct sets of yarns or threads are interlaced at right angles to form a fabric or cloth. Similar methods are knitting, felting and braiding or plaiting. The longitudinal threads are called the warp and the lateral threads are the weft or filling. (Weft or woof is an old English word meaning 'that which is woven'.) The method in which these threads are inter woven affects the characteristics of the cloth.

Cloth is usually woven on a loom, a device that holds the warp threads in place while filling threads are woven through them. A fabric band which meets this definition of cloth (warp threads with a weft thread winding between) can also be made using other methods, including tablet weaving, back-strap, or other techniques without looms. The way the warp and filling threads interlace with each other is called the weave. The majority of woven products are created with one of three basic weaves: plain weave, satin weave, or twill. Woven cloth can be plain, or can be woven in decorative or artistic design.

4.2   Weaving process

The weaving process consists of five basic operations, shedding, picking, beating-up, left off and take up.

Shedding: Separating the warp yarns into two layers by lifting and lowering the shafts, to form a tunnel known as the 'shed'.

Picking or filling: Passing the weft yarn (pick) across the warp threads through the shed.

Beating-up: Pushing the newly inserted weft yarn back into the fell using the reed.

Let off: The warp yarns are unwound from the warp beam during the above three processes.

Take up: The woven fabric is wound on the cloth beam during the above three processes.

The above operations must be synchronised to occur in the correct sequence and not interfere with one another. The full sequence is repeated for the insertion and interlacing of each weft yarn length with the warp yarns and is therefore called 'The Weaving Cycle'.

4.2.1   Shedding mechanisms

All weaving machines control the warp yarns to create a shed. This can be accomplished with the following systems:

1. Crank shedding.
2. Cam shedding or tappet shedding.
3. Dobby shedding.
4. Jacquard shedding.

Crank, cam and dobby mechanisms control the harnesses which lift the shafts. Jacquard machines control the individual warp yarns. Each system is outlined below.

Crank shedding

Crank shedding mechanisms are simple and relatively cheap to use. However it can only be used for plain weave fabric constructions. In this system the harnesses are controlled by the crank shaft of the weaving machine. For each crank shaft revolution a wheel is rotated half a turn, which changes the harness position. This system is only used in air-jet and water-jet machines where high speed is achieved.

Cam shedding

Cam shedding is also simple and inexpensive. A cam is a disk which has grooved or conjugated edges which corresponds to the lifting plan. The lifting plan controls in which harnesses are lifted. The disadvantage of cam shedding is that when the woven design has to be changed the cams have to be rearranged to suit the new design. Pattern design is also limited due to the amount of harnesses the cams can control.

Dobby shedding

Dobby shedding is more complex than crank and cam systems. The main advantage of dobby looms is that more intricate designs can be produced. Older dobby looms were operated by wooden lags with pegs, which rotated around a roller above the loom. The pegs in the lags correspond to the lifting plan, which controls which harnesses are lifted. Punched paper or plastic pattern cards can also be used. Recently modern dobby looms are controlled via an electronic system. The disadvantage of dobby systems is that faults are more likely to occur due to their complexity.

Jacquard shedding

In jacquard weaving a device called a 'jacquard' selects and lifts the warp yarns individually. This type of machine is used for larger more detailed patterns, where all or most of the yarns in a repeat, move independently. There are single or double lift machines which use either mechanical or electronic systems to control the harness lifting and lowering. Modern jacquards are capable of handling over 1200 harness cords which control the lifting and lowering of the warp yarns.

4.2.2   Weft insertion methods

Modern automatic looms do not require a shuttle to carry the weft yarn across the shed. Instead the weft yarn is inserted by either one of the following methods.

Rapier

A shuttleless weaving loom in which the filling yarn is carried through the shed of warp yarns by fingerlike carriers called rapiers.

There are two types of rapiers.

1. A single long rapier that reaches across the loom's width to carry the filling to the other side.
2. Two small rapiers, one on each side. One rapier carries the filling yarn halfway through the shed, where it is met by the other rapier, which carries the filling the rest of the way across the loom. The insertion rate of picks can be up to 1000 m min^{-1}.

Projectile

Projectile machines carry yarn through the shed using a small bullet shaped object known as a 'projectile'. The yarn must be presented to the projectile in order for it to grip this. This process can occur in the following ways:

1. A single projectile is fired from each side of the machine alternately and requires a bilateral yarn supply.
2. A yarn supply from one side of the machine is presented to the projectile. It carries the weft yarn across the machine and is then transported back to the other side by a conveyor belt. Several projectiles are in use at the same time to enable rapid pick insertion. Pick insertion rate can be up to 1300 m min^{-1}.

Air jet

In air-jet weaving machines the filling yarn is inserted pneumatically. It is carried through the shed by compressed air flow supplied from a main nozzle and relay nozzles. This is the fastest type of weaving enabling pick insertion of 3000 m min^{-1}.

Water jet

Water jet weaving is the same principle as air jet weaving, water is used instead of air and a similar speed is achieved. One disadvantage is that only hydrophobic yarns can be used.

Multiphase

All of the above methods are classified as single phase weaving, where by the weft yarn is laid across the full width of the warp yarns and beat-up takes place. Multiphase weaving involves several phases in which weaving taking place at the same time, so that several picks can be inserted simultaneously. The shedding mechanisms of the weaving affect this process:

1. Wave shed machines carry the yarn in either straight or circular paths. Parts of the warp are in different stages of the weaving cycle at any one moment. It is possible for a series of weft carriers to move along in successive sheds in the same plane.

2. In parallel shed machines numerous sheds are formed simultaneously. Each shed extends across the full width of the warp and moves in the warp direction.

Limitations and energy consumption: Although air and water jet machines can weave fabric at higher speeds compared to the projectile and rapier looms, the high power consumption results in higher costs. The flow of the air is also difficult to control and waste heat produced by the compressors is sometimes wasted when it could be used for other operations in the factory. However cooling of the factories via air conditioning is not always necessary with air and water jet looms and so energy costs are saved in this way. The rapier and projectile looms produce a lot more heat and so air conditioning is often installed to keep temperatures down within the factory. Multiphase wave shed looms moving in a straight path have not been commercially successful as maintaining a clean shed has proved very difficult.

4.3 Energy-efficiency technologies and measures in the weaving process

The list of measures/technologies for the weaving process are discussed below.

Evaluation and enhancement of the energy efficiency of compressed air systems in air-jet weaving plants: Air-jet weaving machines use compressed air for weft yarn insertion. The conversion efficiency to produce compressed air is fairly low (less than 15% efficiency without heat recovery). Most textile companies rely on compressed air in their production and improving the use of compressed air will have significant economic benefit for plants.

For example, in a case-study conducted by Georgia Technical Institute, it was found that by reducing air leakage from 12% to 6% and lowering system air pressure by 16 psi, a saving of about US$440,000 annually is expected for a compressed air system operating 500 looms (weaving machines). It was found that loom performance was not significantly affected by air pressure until the pressure drops below a certain level, beyond which a drastic increase in filling stops was observed.

In this study, unacceptable weaving performance only occurred when the pressure in the flowmeter was lower than 64 psi and this represents the minimum pressure required for the experimental loom, a level considerably lower than the manufacturer's suggested pressure of 80 psi.

Replacement of conventional motors with energy efficient motors in ring frames: The motor losses can be categorised as:

1. The resistance loss in winding—copper losses.
2. The magnetic current loss—iron losses.
3. Windage and friction losses.

In energy efficient motors, now available for LT range, all these losses are reduced, resulting in efficiencies of 93–95%.

The following major modifications have been adopted in the motor design, resulting in increase in efficiency.

1. Using more copper in the motor winding reduces copper loss.
2. Better stamping design.
3. Windage and friction losses are reduced by proper selection of bearings having optimum clearance.
4. Better cooling fan design.

Benefits of energy efficient motor: Reduction in electrical consumption.

The specific energy consumption reduced from 0.542 units/kg of yarn to 0.482 units/kg of yarn, for the same count.

Cost benefit analysis:

1. Annual savings   — $20000
2. Investment   — $27000
3. Simple payback   —   16 months

Thus, each plant is unique in its own way and what is applicable in one plant may not be entirely applicable in another identical unit. Hence these case studies could be used as a basis and fine-tuned according to the individual plant requirement before taking up for implementation.

General measures to save energy in weaving plants: Weaving machines (looms) account for about 50–60% of total energy consumption in a weaving plant. Humidification, compressor and lighting accounts for the rest of the energy used, depending on the types of the looms and wet insertion techniques. Since a loom is just one machine, there are not many physical retrofits that can be done on existing looms to improve their efficiency. Of course the looms differ in their energy intensity (energy use per unit of product).

However, for a given type of the loom, most of the opportunities for energy-efficiency improvements are related to the way the loom is used (productivity), the auxiliary utility (humidification, compressed air system, lighting, etc.), and the maintenance of the looms.

All the measures mentioned in the previous sections which improve the efficiency of humidification and compressed air systems used in spinning processes are also to a great extent applicable to weaving plants. In addition to these, the below measures for efficiency improvements of the weaving process are also available opportunities:

1. Loom utilisation should be more than 90%. A 10% drop in utilisation of loom machines will increase specific energy consumption by 3–4%.

2. The type of weaving machine can significantly influence the energy use per unit of product. Therefore, when buying new looms, the energy efficiency of the loom should be kept in mind. However, it should be noted that some looms can only produce fabrics with certain specifications and not all looms can produce all types of fabrics.

3. The quality of warp and weft yarn directly influences the productivity and hence efficiency of the weaving process. Therefore, using yarns with higher quality that may have a higher cost will result in less yarn breakage and stoppage in the weaving process and can eventually be more cost-effective than using cheap, low quality yarns in weaving.

Rationalisation in fabric production is such that while various improvements in machinery aimed at high speed operation and labour saving have been carried out, the amount of energy use per unit of the product has gradually increased. Regarding loom design, high productivity shuttleless looms such as water jet, rapier and gripper types have successfully been introduced, with air jet models put in practice in the production area of industrial fabric material.

The amount of energy consumed by each loom during its weaving operation can be estimated from the motor capacity and weaving speed. Conventional shuttle looms are based on the weft-insertion method, incorporating a shuttle zooming to and fro with a large inertia mass and mounted with extra weft and they also use energy consuming pirns as an integral part of the machine. For this reason, the shuttleless looms contribution to energy saving cannot be regarded as too high.

On the other hand, as large amount of energy is consumed in sizing, as one of the preparatory operations for weaving, the introduction of foam and solvent sizing operations are being investigated. Furthermore, long fibre fabrics using nonsizing filaments have been developed, eliminating the sizing process altogether. In a reported example, the introduction of a new heat exchanger into a sizing machine with a very poor sealing capability achieved more than 40% of energy saving.

Energy conservation in wet processing

5.1   Introduction

Wet processing is one of the major streams in textile processing refers to textile chemicals process engineering and applied science. The other three streams in textile engineering are yarn engineering, fabric engineering and apparel engineering. Wet process is usually done on the manufactured assembly of interlacing fibres, filaments, and/or yarns having substantial surface (planar) area in relation to its thickness and adequate mechanical strength to give it a cohesive structure. In other words, wet process is done on manufactured fabric. The processes of this stream is involved or carried out in aqueous stage and thus it is called wet process which usually covers pretreatment, dyeing, printing and finishing.

Power and utility plays a vital role and their cost contributes significantly on total cost of finished textile product. Being basic and the oldest industry, it grows with the population rate, so is the need of power. Among all the industries, textile sector consumes about 5–8% of the total energy mainly in the form of electrical and thermal. Out of this about 40–45% energy consumed in manufacturing of yarn and fabric and 35–60% energy utilised in wet processing. Textile wet processing involved pretreatment, dyeing, printing and finishing, on grey fibre to impart aesthetic values and marketability. All the four conventional energy sources namely, coal, electricity, oil and gas are utilised in the wet processing of textile. Table 5.1 shows the pattern of steam consumption in a typical composite textile sector.

Table 5.1: Steam consumption in a typical textile composite mill.

Department	Steam consumption (%)
Humidification	10
Sizing	15
Boiler house	05
Wet processing	60
Leakage	10

Wet processing of textile consumes very small proportion of electrical energy mainly for running of machineries. Fuel in terms of coal or oil is used extensively, mainly to generate steam or heat. International energy crisis and escalating cost of fuels have diverted all the researchers and industrialists to think for the ways to conserve energy. Various approaches have been developed and practiced to conserve energy in wet processing namely:

1. Developments of machines with low material to liquor ratio.
2. Efficient heat recovery and processing.
3. Developments in specialty chemicals and dyes to reduce processing time or cycles.
4. Optimise wet pickup on fabric to reduce drying energy.
5. Adoption of e-control to minimise unnecessary leakages.
6. Development of techniques to reduce process cycle.

Innovation of unconventional techniques have opened a new era of energy conservation in textile wet processing. The important benefits to textile industry of the said technologies are:

1. The apparent increase in diffusion rate of chemicals.
2. Energy savings as process operate at lower temperature.
3. Increased efficiency of process leads to less effluent.
4. Preserved drapability, luster and finish of fibre.
5. Overall cost reduction of process.

5.2 Preparatory process

5.2.1 Combine preparatory treatments in wet processing

Combining preparatory treatments such as the combined desizing, scouring and bleaching of a cotton fabric could lead to a process step reduction from the original eight-stage process to just two stages; this method would employ a steam purge and cold-pad-batch technique. The elimination of three intermediate washings, one hot kier and a cold acid process could reduce energy requirements by as much as 80%.

5.2.2 Cold-pad-batch pre-treatment

The cold-pad-batch method can be used for pre-treatment in dyeing and finishing plants. In this method, alkali/hydrogen peroxide is embedded into the fabric using a padder and the fabric is then stored to allow complete reaction between the fabric and chemicals prior to rinse. By using this methods 50% of the water and electricity and 38% of the steam used in pre-treatment can be reduced.

5.2.3 Bleach bath recovery system

Since large amounts of water is used in wet processing operations, recycling and reuse of rinse water can greatly reduce hydraulic loadings in wastewater treatment systems, while at the same time creating savings through reduced water use, energy consumption and wastewater disposal. There are two ways to reuse spent rinse and wash water. This water can be reused in another rinse operation which accepts low-grade rinse water, or it can be reused as process water in wet processing operations with or without the addition of chemicals. Examples include:

1. The reuse of wash water from bleaching in caustic washing and scouring make-up and rinse water.
2. The reuse of rinse water from scouring for desizing or washing printing equipment.
3. The reuse of wash water from mercerising to prepare scouring, bleaching and wetting-out baths.

In these cases, preparation chemicals, especially optical brighteners and tints, must be selected in such a way that reuse does not create quality problems such as spotting. Storage tanks may be necessary to store process liquor for reuse in the makeup of the next bath.

The bleach bath recovery system can be utilised in both the knit and woven segments and can result in a reuse of 50% of the total water used in a typical 100% cotton full bleach process, including pre-scour, bleach and neutralisation. Savings also arise from reduced energy consumption and wastewater disposal charges. The average temperature of the recovered water is estimated to be 40°C. The value of the energy saved has been estimated at US$51,000 per year for a medium size plant.

For example, the warm rinse water in a kier can be used for the next scour liquor, resulting in energy savings of more than 10%.

5.2.4   Use of counter-flow currents for washing

In this system, as the fabric runs through the washing compartments from entry to exit, clean water is passed through the plant from the back to the front. This means that the cleanest fabric comes into contact with the cleanest washing liquor. By applying this counter-flow principle, it is possible to save both water and energy. This process was developed to reduce water consumption for washing and is composed of washing equipment, a washing and dehydration mechanism, filter equipment, sensors and pumps. Washing water is supplied from a direction in the reverse of fabric flow and the sensor detects water impurity to adjust automatically the feed rate of the water.

For example, during dyeing with the cold-pad-batch dyeing method, a padder with nip-controlled rollers is used to apply dyestuffs to the fabric in a defined manner. After a dwelling time (which varies depending on the dyestuff) the excess dyestuff needs to be washed out.

5.2.5 Energy saving in continuous washing machines

Continuous washing machines are generally used after preparatory processes such as– scouring, bleaching and mercerising, or after dyeing. A continuous washing machine (a 'washing range') is made up of a number of tanks, compartments or becks connected by tension compensators and nip rollers. The fabric is threaded, open width, around a series of rollers in each tank. The rollers help to increase liquor agitation and the transfer of impurities, to improve washing efficiency. Figure 5.1 shows a continuous washing range.

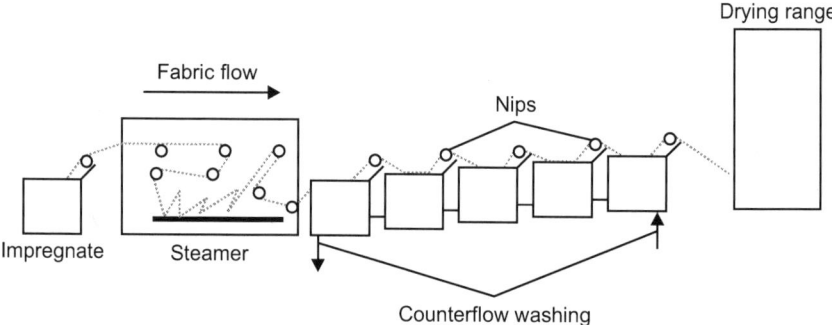

Figure 5.1: Schematic of a continuous washing machine.

5.2.6 Installing covers on nips and tanks in continuous washing machines

The losses at nips are considerable and even in some cases they can exceed 40% of total energy demand of washing machines. Hence, it is important to cover them as well as the hot tanks. Any fitted covers should be easily removable to allow quick access. This simple retrofit measure can result in significant energy savings.

5.2.7 Installing automatic valves in continuous washing machines

Automatic stop valves which link the main drive systems of machines to water flows can save considerable amounts of energy and water by shutting off water flow as soon as a stoppage occurs. A series of shorter stoppages typically accounts for up to 20% of machine working time. Often both the water flow and the heating are left on throughout these short interruptions, which in total will result in significant energy and water wastage. The payback period for installing automatic stop valves on continuous washing machines could be as low as one month.

5.2.8 Installing heat recovery equipment in continuous washing machines

Installing heat recovery equipment on a continuous washer is usually a simple but very effective measure since water inflow and effluent outflow are matched and this eliminates the need for holding tanks. The effluent from these machines can become contaminated with fibrous material, so it is important to install a heat exchanger capable of handling such loads. One option is a self-cleaning, rotating element exchanger which has an efficiency of about 70%. Another measure is to install a simple plate heat exchanger with a pre-filter, which may have a higher initial cost, but which also has an efficiency that could be higher than 90%.

5.2.9 Reduce live steam pressure in continuous washing machines

A reduction in live steam pressure can prevent steam breakthroughs, thus improving heat transfer efficiency in direct steam heating applications. Similarly, reducing steam pressure in closed coils will have the advantage of the lower pressure steam has higher latent heat content.

5.2.10 Introducing point-of-use water heating in continuous washing machines

Point-of-use gas-fired water heaters can be used to enable processes to be run independently of plant central boiler systems. This means that boiler and distribution losses associated with centralised systems (which can be as much as 50% of the fuel input). Point-of-use heating also offers greater flexibility since it allows operation outside of main boiler operating hours. This measure, however, requires significant changes to the washing machine and may have a high capital cost.

5.2.11 Interlocking the running of exhaust hood fans with water tray movement in yarn mercerising machines

In many plants, the exhaust hood fans fitted on the fume extractor hoods of yarn mercerising machines remain on during the working time of the machine. However, the running of exhaust hood fans can be electrically interlocked with the forward movement of the water tray, as the fans have to remove the fumes generated during the washing phase only. The implementation of this measure will result in electricity savings in fan use. One case-study showed electricity savings of 12.3 MWh/year/machine. The investment cost for this measure is minimal.

5.2.12 Energy saving in cooling blower motors by interlocking motors with the fabric gas singeing machine's main motor

It often happens in textile plants that for many machines, while a machine is stopped the auxiliary units are still running continuously, resulting in energy waste. One such case is the operation of fabric singeing machines. In many plants, the cooling blower is running continuously even after the singeing machine is stopped. The solution to this is to interlock the cooling blower motor with the singeing machine's main motor, thereby saving energy. The implementation of this measure has resulted in 2.43 MWh/year/machine electricity savings. The investment cost of this measure is very small.

5.2.13 Energy saving in shearing machine blower motors by interlocking motors with the main motor

Similar to the measure mentioned above for the fabric gas singeing machine, the interlocking of blower motors with the machine's main motor can also be implemented in fabric shearing machines. Based on a case-study of the implementation of by using this measure in a textile plant, energy savings of 2.43 MWh/year/machine can be achieved. The investment cost of this measure is zero or very small.

5.2.14 Enzymatic removal of residual hydrogen peroxide after bleaching

Raw cotton fabric requires bleaching, usually achieved with hydrogen peroxide. To achieve reproducible bleaching results, residual hydrogen peroxide content of 10–15% of the initial quantity should be still available at the end of bleaching. The residual hydrogen content must be completely removed to prevent any change of shade of dyestuffs, which are sensitive to oxidation. Reducing agents and several rinsing steps are necessary in common peroxide removal techniques. High energy and water consumption and the use of sulfurous reducing agents are main disadvantages of the conventional technique.

Special enzymes (peroxides) catalyse the reduction of hydrogen peroxide to oxygen and water. No side reactions with the substrate or with dyestuffs occur. Peroxides are completely biodegradable. Rinsing steps after peroxide bleaching can be reduced with enzymatic peroxide removal (normally only one rinsing step with hot water is necessary). Peroxides have no negative influence on downstream dyeing processes. A typical process include these steps: peroxide bleaching - liquor change - one rinsing bath (hot water) - liquor change - enzymatic peroxide removal - dyeing without previous liquor change. Enzymatic peroxide removal is possible in a discontinuous, semi-continuous and continuous production method. The method is applicable both in new and existing installations. Savings in water and energy consumption can be in the range of 6–8% of production costs. Barclay and Buckley reports that approximately US$15 for 30/T fabric can be saved, depending on the local cost of water and energy. The Skjern Tricotage-Farveri textile plant in Denmark implemented this measure and achieved 2,780 GJ/year in energy savings and 13500 kL/year in water savings.

5.2.15 Enzymatic scouring

Desizing, scouring with strong alkali and bleaching are typical pre-treatment steps in cotton finishing plants. Scouring and bleaching steps are often combined. With the use of enzymes the alkaline scouring process can be replaced. Due to the better bleaching ability of enzyme, scoured textiles bleaching can be carried out with reduced amounts of bleaching chemicals and auxiliaries. With enzymatic scouring, sodium hydroxide used in common scouring is not necessary. In term of quality, the enzymatic scouring will achieve good reproducibility, reduced fibre damages, good dimensional stability of fabrics, increased colour yield and a soft fabric texture.

The rinsing water consumption for scouring can be reduced to 20% of traditional techniques. If the enzymatic scouring is combined with bleaching with reduced concentrations of peroxide alkali, the rinsing water consumption can be reduced to 50% of traditional techniques; hence saving energy. This method can be applied to cellulosic fibres and their blends (woven and knitted goods) in continuous and discontinuous processes. Existing machines (jets, overflows, winches, pad batches, pad steamers and pad rollers) can be used for this purpose. This method is applied world-wide, especially in German finishing plants. The method offers savings in water, time, chemical auxiliaries and energy, depending on on-site conditions.

5.2.16 Use of integrated dirt removal/grease recovery loops in wool scouring plants

Wool scouring is an energy-intensive process. In addition to generally applicable good housekeeping techniques, the biggest energy savings in the wool scouring process come from reducing effluent flowdown to drains or to onsite effluent treatment plants by the installation of a dirt/grease recovery loop.

Techniques include fitting a heat exchanger to recover heat from the dirt/grease loop flowdown. The implementation of dirt removal/grease recovery loops allows:

1. A reduction in water consumption ranging from 25% to more than 50% (the consumption of water of a conventional plant operating in counter-current is between 5 and 10 L/kg of greasy wool).

2. Energy savings from a dirt/grease recovery loop can be estimated at about 2 MJ/kg of greasy wool if a scour with loop and a heat exchanger is used.

3. The production of wool grease which is a valuable by-product.

4. A reduction in detergent consumption proportional to the water savings achieved.

5. A reduction of the load sent to the effluent treatment plant, which means a reduction in the consumption of energy and chemicals for the treatment of the waste water.

5.3 Dyeing and printing process

5.3.1 Installation of variable frequency drives on pump motors of top dyeing machines

Top dyeing is a method for dyeing combed wool before spinning. In this process, the wool 'top' is placed in large vats and dye liquor is circulated through the tops at high temperatures. Variable frequency drives (VFDs) can be installed on the pump motor of the top dyeing machine in order to save energy by setting the speed of the pump motor based on the dyeing process requirements. A textile plant in India installed VFDs on the pump motors of 25 top dyeing machines. The average electricity saving resulting from this retrofit was about 26.9 MWh/year/machine. The investment cost is reported to be about US$3100 for each machine.

5.3.2 Heat insulation of high temperature/high pressure (HT/HP) dyeing machines

Insulation of pipes, valves, tanks and machines is a general principle of good housekeeping practice that should be applied in all steam consuming processes in textile plants. It is reported that insulation can save up to 9% of the total energy required in high temperature/high pressure wet processing machines. The insulation material may be exposed to water, chemicals and physical shock. Any insulation should, therefore, be covered or coated with a hard-wearing, chemical/water resistant outer layer. The fuel saving potential of such insulation in a plant is reported to be as high as 210–280 GJ/year with an investment cost of US$9000–US$13,000 per plant (including both the cost of insulation materials and installation). The payback period is about 3.8 to 4.9 years.

5.3.3 Automated preparation and dispensing of chemicals in dyeing plants

Automated chemical dosing and dispensing systems are now commonly used in many companies in the textile industry. In modern dosing and dispensing systems, the water used for washing the preparation vessel and supply pipes is taken into account when the quantity of liquor to be prepared is calculated. This approach reduces waste water, but still involves the premixing of chemicals. Other automated dosing systems are available where the chemicals are not premixed before being introduced into the applicator or dyeing machine. In this second case, individual streams are used for each of the products. As a result, there is no need to clean the containers, pumps and pipes before the next step, saving even more chemicals, water, energy and time. This is an important feature in continuous processing lines.

5.3.4   Automatic dye machine controllers

Automatic dye machine controllers offer an effective means for enhanced control over dyeing processes. They are based on microprocessors and allow for feedback control of process parameters such as pH, colour and temperature. They analyse process parameters continuously and respond more quickly and accurately than manually controlled systems. Dye machine controllers can be retrofitted for many of the dye machines used in plants.

Automatic dye machine controllers control the dye cycle, including the amount of water utilised in the process and hence the amount of water and pollutants discharged as effluent.

Automatic host controllers include dye programme management and reporting systems, on line scheduling, recipe management and costs analysis (energy, dyes and chemicals, etc.). Dye machine controllers have the potential to reduce the volume of industry effluents by up to 4.3%.

5.3.5   Cooling water recovery in batch dyeing machines (jet, beam, package, hank, jigger and winches)

Cooling water and condensate water are non-process water uses. Many cooling water systems are operated on a once-through basis. In general, cooling water and condensate water can be pumped to hot water storage tanks for reuse in functions where heated water is required, such as for dye makeup water, bleaching, rinsing and cleaning. Recovery of cooling water can save about half of the total energy requirements for dyeing under pressure at high temperatures.

The recovery of cooling water also reduces water and effluent costs. The cooling water contains approximately 43% of the total energy input. Much of this energy can be recovered by directing the cooling water stream into a hot water storage system for reuse at 50°C–60°C. However, care must be taken to restrict the initial dye liquor temperature to avoid dyestuff strike-rate problems and fabric creasing.

5.3.6 Discontinuous dyeing with airflow dyeing machines

Discontinuous processing of textile products, in general, requires more water and energy compared to continuous processes. For a long time, efforts were undertaken to optimise discontinuous processes with respect to productivity, efficiency and also to minimise energy and water consumption. This has led to the development of jet dyeing machines in which liquor ratios have been reduced step by step. The latest developments (airflow dyeing machines) have a liquor ratio of 1:3 (for woven and polyester fabrics) and 1:4.5 liquor ratio (for woven cotton fabrics), instead of 1:10–1:12 in conventional jet dyeing machines. Additionally, airflow dyeing machines offer a combination of high productivity and reduced water, chemicals and energy consumption. To achieve such low liquor ratios, within the jet dyeing machine the fabric is moved by moisturised air or a mixture of steam and air only (no liquids), aided by a winch. The prepared solutions of the dyestuffs, auxiliaries and basic chemicals are injected into the gas stream.

The application of this technology requires investments in new dyeing machines as existing machines cannot be retrofitted. These machines can be used for both knit and woven fabric, however this process is not applicable to fabrics consisting of wool or wool blends with a percentage of more than 50% wool because of felting problems. Due to lower liquor ratios, new airflow-dyeing machines are expected to save up to 60% of both water and heating energy, 40% in chemicals, 35% in salts and even 10% in dyestuffs.

5.3.7 Installation of VFD on circulation pumps and colour tank stirrers

Circulation pumps are used to circulate chemicals in machine chambers in the dye house. In many plants, especially those with old equipment, the flow of chemicals is often controlled by closing ball valves. VFDs can be installed instead of ball valves for flow control, thereby saving energy. Stirrers are installed on colour tanks and are used for mixing the colours. VFDs could also be installed on the stirrers to control their speed, as it is not necessary to run the stirrers at full speed. The implementation of these two retrofit measures in a plant in India resulted in 138 MWh/year electricity saving. The investment cost associated with this saving was about US$2300.

5.3.8 Dyebath reuse

Dyebath reuse is the process by which exhausted hot dyebaths are analysed for residual colorant concentrations, replenished and reused to dye additional batches of material. Dyebath reuse reduces effluent volume and pollutant concentrations in the effluent. Dyebath reuse carries the higher risk of shade variation because impurities can build up in the dyebath and decrease the reliability of the process. If properly controlled, dyebaths can be reused for 15 or more cycles (ranging from five to 25 cycles). For maximum dyebath reuse benefits, dye classes that undergo minimal changes during the dyeing process (such as acid dyes for nylon and wool, basic dyes for acrylic and some copolymers, direct dyes for cotton and disperse dyes for synthetic polymers) must be used. Vat, sulphur and fibre reactive dyes are very difficult to reuse. One complication is that each of the individual colour components added to the dye bath to achieve the desired outcome colour may have different affinities for binding to the fabric; thus, replenishing the bath correctly may require that different proportions of individual colours to be added compared to the initial recipe. This adds to the complexity of getting the colour right for reuse.

Dyebath reuse offers a return on the investment in the form of dye, chemical and energy savings and the reduction of waste water and waste water treatment. Typical costs for dyebath reuse include capital costs of about US$24,000 to US$34,000 per dye machine for lab equipment and machine modifications and additional annual operating costs of $1,000 to $2,000 per dye machine. Typical annual savings (in the form of dyes, chemicals, water, sewer and energy reductions) is about US$21,000 per dye machine.

5.3.9 Equipment optimisation in winch beck dyeing machines

A number of technological improvements have been introduced in winch beck dyeing machines, which are:

1. Heating: The liquor in the early winches was usually heated by direct steam injection through a perforated pipe. This system provided both rapid heating and vigorous agitation in the beck, but entailed dilution which had to be taken into account. Indirect heating/cooling is now more commonly used to overcome dilution and water spillage.

2. Liquor ratio: Recently developed winch becks operate at liquor ratios that are significantly lower than conventional machines. Moreover, an outstanding feature is that small batches can be dyed with approximately the same liquor ratio as for maximum loads.

3. Rinsing: Modern winches are designed to remove the carpet without draining the bath and without cooling or diluting the bath with rinsing water. This system is called the 'hot-drawing-out system', in it the carpet is automatically taken out of the beck and passed over a vacuum extractor which removes the non-bound water. The recovered liquor is diverted back to the dye bath.

4. Modern winches are fitted with hoods to help maintain temperature and minimise losses.

5. Modern winches are also equipped with automated dosing and process control systems for full control of the temperature profile and the chemicals injected during the dyeing process.

The features described above bring about substantial savings in terms of water, chemicals and energy consumption. Reductions of 40 to 50% in fresh water for the total dyeing process (up to 94% of these savings occurring in rinsing water reductions) and a 30% reduction in electricity consumption are claimed by the machines' manufacturers.

The advanced concept applied in this type of winch beck is that the rinsing step is no longer carried out in batches, but rather in a continuous mode in a separate section of the equipment without contact between the substrate and the bath. Due to this method, there is no mixing between the rinsing water stream and the hot exhausted bath water, which makes it possible to reuse both streams and to recover the thermal energy.

5.3.10 Reducing the process temperature in wet batch pressure-dyeing machines

A reduction in the process temperature may also be achieved in wet batch pressure-dyeing machines by introducing alternative processes. For example, under suitable circumstances, direct dyeing machines operated at 100–120°C may be replaced with reactive dyeing at 40–60°C, thus minimising water heating and radiation/convection losses. The application of this measure, however, should be assessed in plant as well as on a product basis.

5.3.11 Use of steam coils instead of direct steam heating in batch dyeing machines (winch and jigger)

In older batch dyeing machines like winches and jiggers, dyebaths are traditionally heated by sparging with raw steam. This is a very inefficient use of steam for heating the dyebath. A steam coil submerged in the dyebath now allows for the recycling of the condensate, resulting in significant fuel savings. A plant in Canada replaced their steam sparging system with a steam coil for dyebath heating and achieved fuel savings of 4580 GJ/year. The cost of the replacement is reported to be about US$165,500.

5.3.12 Reducing the process time in wet batch pressure-dyeing machines

Processing times can sometimes be reduced simply by making modifications to the temperature profiles of certain dyeing cycles. This can result in energy savings and improved productivity. Preparatory processes can also be sped up just by the use of different chemical formulations. One example, which involved modifying a kier scour/bleach formulation, is reported to have reduced processing times from ten hours to just over two hours. This measure, however, is not applicable to all processes and its potential application should be studied on plant/process and product bases.

5.3.13   Installation of covers or hoods in atmospheric wet batch machines

Using covers or hoods in atmospheric wet batch machines may seem obvious, but many jiggers and winches are operated at high temperatures without hoods or with open hoods. Using covers or hoods can reduce evaporative losses by approximately half. Evaporation is particularly important in processes with temperatures above 60°C.

5.3.14   Careful control of temperature in atmospheric wet batch machines

Overheating is a common problem in atmospheric wet batch machines. It is most often caused by poor controls, especially in older machines. The maximum achievable temperature in an atmospheric vessel is 95–100°C. Once the dye liquor is boiling, further heat input will not raise the temperature, but will increase evaporation. Although a faster boil does lead to greater agitation of the fabric, this can be achieved more efficiently by installing a circulator. At temperatures above 80°C, live steam breakthrough may occur; as much as 15% of steam can be lost in this way.

5.3.15   Jiggers with a variable liquor ratio

In textile processing, the delivery of relatively small batches of textile products is becoming increasingly important. In some production facilities, more than 50% of the overall production volume consists of such small batches. Conventional jiggers have a number of disadvantages for these jobs. They do not have a variable liquor ratio, which is why the quantities of water, pigments and chemicals cannot be adjusted properly to the varying quantities of fabric being processed. Additionally, starting up a new batch requires prolonged cooling of the equipment, the washing of the fabric requires a lot of time and extra fabric has to be added at the end of a batch to prevent the fabric from slipping off the presses.

In order to cope with these disadvantages, a new generation of jiggers with a variable liquor ratio has been developed. These jiggers make use of a heat exchanger, allowing the heat to be removed and applied elsewhere in the plant. In each passage, the length of the cloth is measured, so extra fabric at the end of the batch can be avoided. Because of additional features such as a vacuum system and sprinklers, the number of passages in washing cycles can be reduced significantly (in a specific case, from twenty to five cycles). In one particular case, energy savings of 26%, water savings of 19% and a 5% reduction in the use of chemicals were achieved.

5.3.16 Heat recovery of hot waste water in autoclaves

Autoclave (high temperature/high pressure) dyeing machines generate relatively high temperature waste water at 75°C, which in many plants is wasted away directly through drain disposal. On the other hand, fresh water at 13 to 25°C is heated to 130°C in the steam heater. A heat exchanger and surrounding equipment like water tanks and pumps for recovering heat from hot waste water as a heat source can be installed. A plate-type heat exchanger is usually recommended. Steam condensate can be recovered in hot water tanks. Waste water coming from autoclave dyeing machines as well as other machines like continuous washers can be treated in the same heat recovery system. By implementation of this measure in a textile plant in Iran showed a fuel savings of 554 MJ/batch in autoclave machines can be achieved.

5.3.17 Insulation of uninsulated surface of autoclaves

If the surface of the autoclave is not insulated, the surface temperature is as high as 100°C to 110°C. This does not only waste energy but also makes for very unpleasant working conditions. Hence, all the hot surfaces should be insulated, including those of the main vessel, air vent tank, heat exchanger and water circulation piping. Water-resistant, easy-paste type insulation material is usually recommended. By using insulation calculation software, energy savings of 15 MJ/batch have been reported.

5.3.18 Reducing the need for reprocessing in dyeing

One of the main causes of reprocessing is the difficulty and time-consuming nature of fabric sampling procedures, especially on older machines. It is therefore vital that dyehouses aim to achieve correct shades quickly and consistently. This may be done simply by improving manual control through better staff training. For large installations there are complete dyehouse management and control systems which are capable of real time machinery supervision, dye cycle editing and production scheduling.

Dyehouse control systems are gradually shifting away from the more rigid read-only memory programmes, to flexible software programming which enables schedules and processes to be tailored on-site. Product quality and productivity can be improved while the use of dyes, chemicals, water and energy are optimised. Improved controls will typically lead to just 5% of the product requiring shading, which results in energy savings of around 10–12%.

5.3.19　Recovering heat from hot rinse water

In textile wet processing, large amounts of hot water (up to 80°C) is used to rinse fabric or yarn. Plants may discharge a mass of rinse water up to thirty times the weight of the yarn/fabric that is rinsed. The heat from the rinse water can be captured and used for pre-heating the incoming water for the next hot rinse. This option provides the important ancillary benefit of reducing the temperature of the wastewater prior to treatment as well.

However, many textile plants around the world are not recovering heat from their hot rinse water.

Plate heat exchangers can be used to transfer the heat from the rinse wastewater to the incoming cold freshwater. Simple heat exchangers are sufficient for continuous processes. In discontinuous processes, the heat exchanger would have to be fitted with buffer tanks and process control devices. Based on case-studies in several textile plants in China, NRDC (Natural Resources Defense Council) reported fuel savings of 1.4–7.5 GJ/T fabric rinsed by the implementation of this measure. The investment cost is reported to be between US$44,000 and US$95,000 depending on mill size and layout. The payback period was less than six months in all cases studied by NRDC.

5.3.20　Reuse of washing and rinsing water

Generally, the rinse water resulting after bleaching or discontinuous dyeing can be reused several times. The rinse water of bleaching processes can often be reused for rinsing after a caustic treatment. In some cases, the water can be reused for a third time to rinse degraded sizing agents. After discontinuous dyeing, the final-step rinse water is hardly contaminated and can possibly be reused for the first rinsing step of the next dyeing process. If the fabric at the start of the process is put into the bath and is taken out after washing, then the rinse water can remain in the bath and there is no need to store it separately.

The wash water can also be used in pre-treatment before the dyeing process. In practice, the reuse of rinse water from dyeing processes is complicated due to pigments remaining in the water (especially in the case of dark colours). Consequently, a prerequisite for recycling is the application of light colours (low pigment concentrations) of pigments with a high fixation rate. Before implementing this measure in a specific company, an assessment has to be made of whether the wash and/or rinse water is sufficiently clean to be reused.

Energy savings per washing machine by the implementation of this measure are significant and annual water savings of 3000 m^3–8000 m^3 can be achieved. Recycling of the last rinse water requires only a small capital investment for collection, pumps and a filter or sieve to capture contaminants. Further reutilisation that requires purifying the water requires a larger investment.

5.3.21 Reduce rinse water temperatures

Rinse water for rinsing after dyeing is heated to a temperature of about 60°C to produce a good quality product. Operational practices in different plants have demonstrated that a reduced temperature of about 50°C can be used without degrading product quality. This will result in significant fuel saving. A plant in Georgia, U.S., has reported a 10% reduction in fuel use by implementing this measure. There is no investment cost for the implementation of this measure. Hence, it will result in substantial cost savings without upfront investment.

5.4 General energy-efficiency measures for wet processing

5.4.1 Automatic steam control valves in the desizing, dyeing and finishing of denim fabric

Steam is extensively used in textile wet processing, such as in the desizing, dyeing and finishing of fabrics. In many plants, especially old ones, steam use is controlled manually in many parts of the process. This usually results in a significant waste of steam. A retrofit programme can be introduced in such plants to replace the manual steam control system with an automatic system. An automatic steam control system controls the supply of the steam to each piece of equipment based on the requirements of the process, which are pre-set in the system; thereby avoiding an excess supply of steam into the machines. A denim fabric manufacturing plant in India installed automatic steam control valves in all desizing, dyeing and finishing processes and has reported 3250 GJ/year in fuel savings.

The capital cost associated with these savings is reported to be about US$5100 for this case-study. The energy savings and cost can vary on a plant basis depending on the number of equipment on which the automatic steam control valves are installed and the operation of the plant.

5.4.2 Recovery of condensate in wet processing plants

Textile plants rely on a large amount of saturated steam in dyeing and other wet processes. Some of that steam converts into condensed water (condensate) over the course of its use. This condensate has a high temperature and purity. One of the best places to collect large volumes of condensate which forms in the woven fabric dyeing and finishing plants is in the drying cylinders, where fabric is dried by making contact with hot cylinders heated by steam. Knitted fabric plants find large sources of condensate primarily in steam traps. The most efficient use of condensate is to return it to the boiler and convert it back into new steam. However, for companies that buy their steam from an outside supplier or whose boiler is located too far from the process in which steam is used, condensate can serve as a water supply for washing or desizing, thereby recovering both water and heat.

5.4.3 Heat recovery from the air compressors for use in drying woven nylon nets

Depending on compressor type and loading, up to 85–90% of the electricity drawn by an air compressor is eventually wasted as heat. With simple safety modifications, waste heat from air compressors can be used for space heating. The alternative is to use the waste heat in a process where low heat is required. In a U.K. based textile plant that produces woven nylon nets, a system was designed to use the waste heat from compressors in the net drying systems.

The plant designed and built a simple heat recovery system. The waste heat from the compressor is ducted into a converted shipping container. Inside the container, the waste heat is drawn through a bank of trolleys on which the wet nets are arranged. The hot dry air is drawn through the nets by an electric fan fitted to the base of each trolley, absorbing moisture along the way. The moist air stream is discharged to the atmosphere through vents cut into the side of the container.

5.4.4 Utilisation of heat exchangers for heat recovery from wet process wastewater

Large volumes of heated water are presently being used in textile plants for rinsing in the desizing, scouring and bleaching steps of continuous preparation ranges as well as in dyeing machines. Much of the hot process water and some of the chemicals that are presently being discharged as waste can be recovered and reused cost-effectively. A large number of different techniques for recovering waste heat exist which can reduce process operation cost and conserve significant amounts of fuel.

Energy conservation in drying process

6.1 Introduction

Drying is done after dewatering of fabric. In textile finishing unit; dryer uses for dry the knit, woven fabrics and dyed yarn. But the drying process and drying mechanism of yarn and fabrics is different from one to another. The main functions of a textile dryer is to dry the textile fabrics. Drying is defined as a process where the liquid portion of the solution is evaporated from the fabric. Drying happens when liquid is vapourised from a product by the application of heat. Heat may be supplied by convection (direct dryers), by conduction (contact or indirect dryers), radiation or by placing the wet material in a microwave or radio frequency electromagnetic field. Over 85% of industrial dryers are of the convective type with hot air or direct combustion gases as the drying medium. Over 99% of the applications involve removal of water.

This is one of the most energy-intensive unit operations due to the high latent heat of vapourisation and the inherent inefficiency of using hot air as the (most common) drying medium.

6.2 Drying in textile industry

The final washing-off is followed by drying. This latter stage is usually carried out on cylinders and varies in energy requirement between 1.9 and 3.7 MJ/kg. This variation is, in part, a reflection of variation in fabric width, weight and moisture content.

This sequence of preparation stages converts an inherently coloured, unwettable material into an absorbent, clean and white fabric ready to be dyed or printed or finished. Overall, the preparation process makes a significant contribution to the energy consumption in a dyeing and finishing works where cotton or polyester/cotton fabrics are processed. As stated above, about 30 MJ/kg of energy are used in processing such fabrics. Water consumption is also significant. Typical energy consumption figures for preparatory processes are given in Table 6.1.

It is clear from this information that modifications in the separate stages, resulting in reduced energy consumption, are highly desirable, provided production rates and fabric quality do not suffer. Attempts have been made to combine various processing operations into a single-stage process and also to carry out the preparatory processes at low temperature.

Table 6.1: Energy use in typical preparation processes.

Energy-consuming process	Energy MJ/kg	Steam kg/kg fabric
Scour (pad-steam)	6.2	2.25
Wash-off	5.5	2.00
Bleach (pad-steam)	6.2	2.25
Wash-off	5.5	2.00
Dry	1.9	0.70
Total	25.3	9.20
Rope-kier scour, bleach and wash-off	7.0	2.55
Dry	3.7	1.35
Total	10.7	3.90
J-box scour, bleach and wash-off	8.1	3.00
Dry	2.0	0.70
Total	10.1	3.70

Drying curve: The Fig. 6.1 is a typical curve, in the initial period, drying rate is high; later, drying rate decrease because of interference of free moisture on the surface of the product.

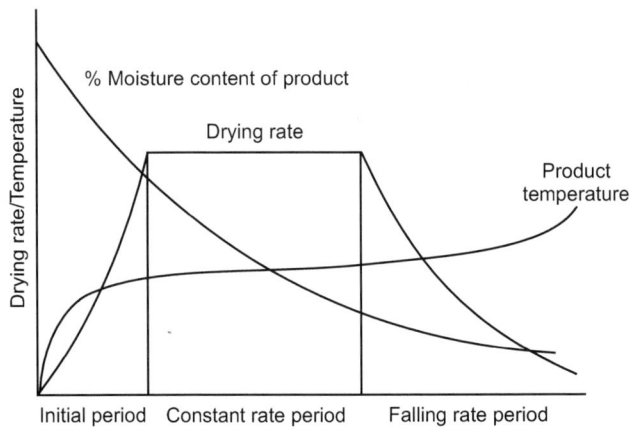

Figure 6.1: Typical curve of drying process.

6.3 Type of dryers used in textile industry

6.3.1 Hot air dryer-stentor

Fabric drying is usually carried out on either drying cylinders (intermediate drying) or on stenters (final drying). Drying cylinders (Fig. 6.2) are basically

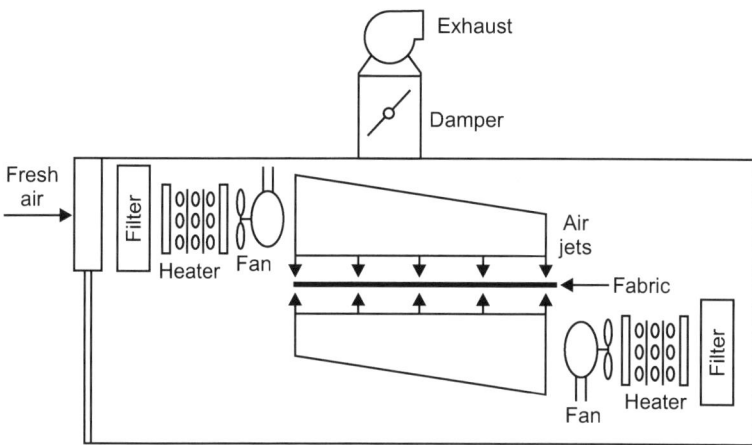

Figure 6.2: Drying clyinders.

a series of steam-heated drums over which the fabric passes. It has the drawback of pulling the fabric and effectively reducing its width. For this reason it tends to be used for intermediate drying. The stenter is a gas fired oven, with the fabric passing through on a chain drive, held in place by either clips or pins. Air is circulated above and below the fabric, before being exhausted to atmosphere. In drying processes, the stenter is used for pulling fabric to width, chemical finishing and heat setting and curing.

Drum dryer used in textile industry

The drum dryer is the best textile dryer. The problem arising from its use is the problem of 'hand' of fabric; after treatment in a hot drum, the fabric becomes rough. This phenomenon is easily overgone; it is due to the hairiness of the fabric. For instance, after a treatment in a domestic tumbler, the hairiness of the fabric is highly developed; the fabric is soft. After a drum drying on a thermal treatment with contact upon a drum, the hairiness is ironed; the fabric cannot be easily compressed and seems rough. To avoid or destroy this roughness, a single brushing is sufficient.

The drum dryer is economical as the airflow is not important: the heat is transferred by contact and the speed of air has no influence. But in a drum dryer, we need a certain quantity of air to evacuate water/steam. Furthermore, some air blown into the pocket increases the production by about 20%.

6.3.2   Contact drying-steam cylinders/cans

This is the simplest and cheapest mode of drying woven fabrics. It is mainly used for intermediate drying rather than final drying (since there is no means of controlling fabric width) and for predrying prior to stentering. Fabric is passed around a series of steam heated cylinders (Fig. 6.3) using steam at pressures varying from 35 psi to 65 psi. Cylinders can be used to dry down a wide range of fabrics, but it does give a finish similar to an iron and is therefore unsuitable where a surface effect is present or required.

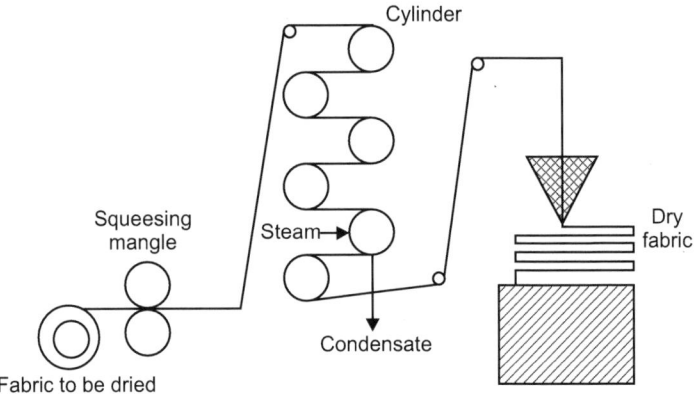

Figure 6.3: Steam heated cylinders.

High frequency (HF) drying

High frequency drying has to be used only for bulk materials. For example, for yarn packages, high frequency dryers are commonly used. But, from the energy viewpoint, a high frequency generator is never a good solution because the efficiency of a high frequency generator is never higher than 55–60%. This efficiency cannot be increased as the energy lost corresponds to the kinetic energy of the electrons. As the price of electricity is about 3 times that of thermal energy, the cost of 1 kWh is 6 times the price of 1 kWh steam. But for a package dryer, if we have to use a classical so called rapid dryer, the packages are brought into the vessel without centrifuging. Then, we have to vapourise 120% water. If the clarinets have to be emptied to dry the packages in a HF dryer, then centrifuging can be done before drying. A good centrifuging eliminates 2/3 of the water. Consequently, HF drying partly avoids the overdrying necessity and with hot air it seems to be a good technique.

6.4 Energy conservation in drying of textile industry

Some of the steps adopted in energy conservation in drying operation are discussed below.

6.4.1 Energy-efficiency improvements in cylinder dryer

Contact drying is mainly used for intermediate drying, rather than final drying (since there is no way of controlling fabric width) and for pre-drying prior to stentering. Fabric is passed around a series of cylinders, which are heated by steam supplied at pressures varying from 35 psi to 65 psi. Cylinders can be used to dry a wide range of fabrics.

In this example the amount of water removed amounts to 50% moisture content by weight of fabric and evaporative losses account for more than 50% of the energy required. Fabric is often dried from over 100% moisture content to completely dry, increasing evaporative losses to around 75% of the total. Some of the energy-efficiency measures that can be implemented on cylinder dryers are explained below.

Mechanical pre-drying in textile industry

Mechanical pre-drying methods such as mangling, centrifugal drying, suction slot or air knife de-watering are used to reduce drying costs by removing some of the water from the fabric prior to contact drying in cylinder dryer. A slot is three times more energy intensive than a typical mangle, but consistently provides lower water retention rates over a range of fabric types. The effectiveness of mangling depends on a number of factors such as: the diameter and hardness of the bowl; the pressure applied; the temperature of the water in the fabric; and the fabric speed. Centrifugal de-watering may also be used for some fabrics, although its tendency to cause creasing means that the process is mainly used to de-water yarn or staple. In terms of cost and performance, centrifuges fall between mangles and suction slots.

Suction slots can be located in the front of a dryer. The slots draw air through the fabric which runs at speed over a slot. The extracted air/water is then filtered and passed through a water separator. Although they are very effective, slots require a high electrical power input (up to about 50 kW). Typical proportionate limit retentions for a variety of fibers, using mangles and suction slots, are shown in Table 6.2.

Table 6.2: Typical retention limits for a number of fibres.

Fibre	Mangle retention (%)	Suction slot retention (%)
Cotton	45–70	40–55
Viscose	60–100	60–80
Diacetate	40–50	27–40
Nylon 6.6	20–40	14–30
Polyester	20–30	10–16
Wool	58–60	35–55

Table 6.2 indicates that, in general, lower retention rates are achieved by suction slots. This is particularly true when suction slots are used to de-water *hydrophobic* fibers (e.g., diacetate, nylon 6.6 and polyester). In practice, the

figures given for mangling are seldom achieved and performance can be quite poor. For example, it is quite common to see retentions of only 80–100% for cotton. This makes the suction slot seem attractive even for the *hydrophilic* fibers (e.g., cotton and viscose), but its relative energy consumption must be taken into consideration.

Selection of hybrid systems in cylinder dryer

The performance of steam cylinders can be enhanced by the use of directed air, either at ambient or elevated temperatures. Directed air equipment helps to disperse evaporated moisture. One example is the Rapidry system, an Indian development, which uses air jets and claims increased drying rates of around 25–30%. The other example is the Shirley Hood which is used for sizing and coating operations. It is reported that Shirley Hoods could increase drying rates by as much as 40%.

Recover condensate and flash steam in cylinder dryer

Since a large amount of steam is used in cylinder dryer, there is also a significant amount of condensate that should be recovered and returned to the boiler house. In addition, flash steam which is produced when condensate is reduced to atmospheric pressure can be recovered as low-pressure steam and used to heat water or other low-pressure steam processes.

End panel insulation in cylinder dryer

The insulation of end sections can reduce heat waste, thereby saving fuel. This measure, however, is more practical for cylinders with a diameter of one meter or more. For small diameter cylinders, this may not be practicable since the steam pipe, condensate pipe and safety valve get in the way.

Select processes for their low water

This measure is looking one step back before drying to see whether the pre-drying processes could be modified or replaced to minimise the amount of water introduced to the fabric. The use of finishes processes using foam, lick rollers or spray application methods are the options that could be considered.

Avoid intermediate drying in cylinder dryer

There are systems which allow finishes to be applied 'wet on wet' to avoid intermediate drying between processes. Typically a fabric is dried two to three times (sometimes even more) during its passage through a finishing process. In addition to being energy intensive, drying is a bottleneck in the finishing process. Hence, elimination of just one step in drying would result in both energy efficiency and productivity increases.

Avoid overdrying in cylinder dryer

Overdrying of fabric is a very common problem in the textile industry. Fibres have an equilibrium moisture regain, or natural moisture level, below which it is useless to dry them. For some fibres the moisture regain value can be quite high. It is therefore important to control the speed of the drying cylinders so that the equilibrium moisture level is not exceeded. Hand-held moisture meters can be used with a roller sensor to monitor the moisture content of fabrics leaving the drying cylinders.

Reduce idling times and using multiple fabric drying in cylinder dryer

Careful scheduling of fabric batches arriving at the cylinders can reduce idling time, thereby saving energy. Similarly, efficiency can be improved by making cylinders extra wide to allow two batches of narrow fabric to run side by side. To maximise energy savings, it is better to have batches with the same length of fabric, so that they start and finish together. This will help to avoid the idling of one side of the cylinders.

Operating cylinders at higher steam pressures in cylinder dryer

Cylinders can be operated at higher steam pressures and temperatures to reduce radiation and convection losses.

Use of radio frequency dryers for drying acrylic yarn

The steam heated dryer, which is used to dry dyed acrylic yarn skeins, can be replaced by a radio frequency dryer. The RF dryer can reduce yarn drying energy costs as well as maintenance and labour costs. It is also found that radio frequency drying acts on the molecular activity in acrylic yarns, causing them to swell or bulk. This bulking makes a more attractive carpet with a better bloom, improving the quality of products. The case study of the installation of RF dryers in place of steam heated dryers in a carpet producing plant for drying the acrylic yarns used in carpet production has been reported to be around a US$45,000 saving in energy cost with a capital investment of US$200,000.

Use of low pressure microwave drying machines for bobbin drying instead of dry-steam heater

Conventionally, bobbin products are dried by the hot-air drying method using a dry-steam heater. Energy savings could be accomplished by switching from dry-steam heaters to low pressure (LP) microwave drying. The LP microwave drying method features good drying efficiency and the capability to prevent products from over-drying, which happens often in hot-air dryers. The implementation of this measure in Japan resulted in 107 kWh/T yarn electricity saving with an investment cost of US$500,000. The payback period of the project in Japan was less than 3 years.

High-frequency reduced-pressure dryers for bobbin drying after the dyeing process

This equipment is a high-frequency reduced-pressure dryer employed in the bobbin drying after the bobbin dyeing process and achieves 20% electricity saving compared with conventional dry steam-type hot air dryers. A change in the method of temperature control in the drying process from fixed temperature controls to programmed temperature controls and optimised control of the temperature of the drying vessel in accordance with the material and quantity permits a major reduction in electricity. This equipment increases the number of pressure switches and employs a two-stage system of setting pressure in which the initial pressure is lowered, eliminating non-uniform drying and shortening the drying process.

Energy conservation in textile finishing

7.1 Introduction

Textile finishing is a process used in manufacturing of fibre, fabric, or clothing. In order to impart the required functional properties to the fibre or fabric, it is customary to subject the material to different type of physical and chemical treatments. For example wash and wear finish for a cotton fabric is necessary to make it crease free or wrinkle free. In a similar way, mercerising, singeing, flame retardant, water repellent, water proof, antistatic finish, peach finish, etc., are some of the important finishes applied to textile fabric.

Broadly it can be classified into following classes,which are used individually or in combination with each other.

7.2 Mechanical finishing in textile industry

Mechanical finishing involves the application of physical principles such as friction, temperature, pressure, tension and many others.

Calendering: A process of passing cloth between rollers (or 'calendars'), usually under carefully controlled heat and pressure, to produce a variety of surface textures or effects in fabric such as compact, smooth, supple, flat and glazed. The process involves passing fabric through a calendar in which a highly polished, usually heated, steel bowl rotates at a higher surface speed than the softer (e.g., cotton or paper packed) bowl against which it works, thus producing a glaze on the face of the fabric that is in contact with the steel bowl. The friction ratio is the ratio of the peripheral speed of the faster steel bowl to that of the slower bowl and is normally in the range of 1.5 to 3.0. The normal woven fabric surface is not flat, particularly in ordinary quality plain weave fabrics, because of the round shape of the yarns and interlacings of warp and weft at right angles to each other.

In such fabrics it is more often seen that even when the fabric is quite regular, it is not flat. During calendering, the yarns in the fabric are squashed into a flattened elliptical shape; the intersections are made to close-up between the yarns. This causes the fabric surface to become flat and compact. The improved planeness of surface in turn improves the glaze of the fabric. The calender machines may have several rollers, some of which can be heated and varied in speed, so that in addition to pressure a polishing action can be exerted to increase lustre.

Compacting: Durable finish imparted on man-made fibres and knitted fabrics by employing heat and pressure to shrink them to produce a crêpey and bulky texture.

Embossing: This particular type of calendering process allows engraving a simple pattern on the fabric.To produce a pattern in relief by passing fabric through a calendar in which a heated metal bowl engraved with the pattern works against a relatively soft bowl, built up of compressed paper or cotton on a metal centre.

Sueding: This process is carried out by means of a roller coated with abrasive material. The fabric has a much softer hand and an improved insulating effect thanks to the fibre end pulled out of the fabric surface.

Raising or napping: The raising of the fibre on the face of the goods by means of teasels or rollers covered with card clothing (steel wires) that are about one inch in height. Action by either method raises the protruding fibres and causes the finished fabric to provide greater warmth to the wearer, makes the cloth more compact, causes the fabric to become softer in hand or smoother in feel; increase durability and covers the minute areas between the interlacings of the warp and the filling. Napped fabrics include blankets, flannel, unfinished worsted and several types of coatings and some dress goods. Other names for napping are gigging, genapping, teaseled, raised.

Wool glazing: This is done on a special machine, which is used to perform functional finishing on wool fabrics after raising.

Shearing: Shearing is an important preparatory stage in the processing of cotton cloth. The objective of 'shearing' is to remove fibres and loose threads from the surface of the fabric, thus improving surface finish.

Stabilisation: A term usually referring to fabrics in which the dimensions have been set by a suitable pre-shrinking operation.

Decating: Also called decatizing. A finishing process applied to fabrics to set the material, enhance lustre and improve the hand. Fabric wound onto a perforated roller is immersed in hot water or has steam blown through it.

Steaming and heat setting: It is done by using high temperatures to stabilise fabrics containing polyester, nylon, or triacetate but not effective on cotton or rayon. It may be performed in fabric form or garment form it may cause shade variation from side-to-side if done prior to dyeing; may change the shade if done after dyeing.

Sanforising or pre-shrinking: Sanforising is a process where by the fabric is run through a sanforiser; a machine that has drums filled with hot steam. This process is done to control the shrinkage of the fabric.The fabric is given an optimum dimensional stability by applying mechanical forces and water vapour.

Fulling: The structure, bulk and shrinkage of wool are modified by applying heat combined with friction and compression.

7.3 Chemical finishing in textile industry

The finishing is applied by means of chemicals of different origins on fabric by mechanical means is discussed below.

Softening: Softening is carried out when the softness characteristics of a certain fabric has to be improved, by always carefully considering the composition and properties of the substrate.

Elastomeric finishes: Elastomeric finishes are also referred to as stretch or elastic finishes and are particularly important for knitwear. These finishes are currently achieved only with silicone-based products.

Crease resistant or crease proofing: Crease resistant finishes are applied to cellulose fibres (cotton, linen and rayon) that wrinkle easily. Permanent press fabrics have crease resistant finishes that resist wrinkling and also help to maintain creases and pleats throughout wearing and cleaning.

Soil release finishes: These finishes attract water to the surface of fibres during cleaning and help remove soil.

Flame retardant treatment: These are applied to combustible fabrics used in children's sleepwear, carpets and curtains and prevent highly flammable textiles from bursting into flame.

Peach finish: In peach finishing the fabric is subjected to (either cotton or its synthetic blends) to emery wheels, makes the surface velvet like. This is a special finish mostly used in garments.

Anti pilling: Pilling is a phenomenon exhibited by fabrics formed from spun yarns (yarns made from staple fibres). Pills are masses of tangled fibres that appear on fabric surfaces during wear or laundering. Loose fibres are pulled from yarns and are formed into spherical balls by the frictional forces of abrasion. These balls of tangled fibres are held to the fabric surface by longer fibres called anchor fibres.

Anti pilling finish reduces the forming of pills on fabrics and knitted products made from yarns with a synthetic-fibre content, which are inclined to pilling by their considerable strength, flexibility and resistance to impact. Anti pilling finish is based on the use of chemical treatments which aim to suppress the ability of fibres to slacken and also to reduce the mechanical resistance of synthetic fibre.

Non slip finish: A finish applied to a yarn to make it resistant to slipping and sliding when in contact with another yarn. The main effect of non-slip finishes is to increase the adhesion between fibres and yarns regardless of fabric construction, the generic term for these finishes would be fibre and yarn bonding finishes. Other terms that can be used include anti-slip, non-shift and slip-proofing finishes.

Stain and soil resistant finishes: Stain and soil resistant finishes prevent soil and stains from being attracted to fabrics. Such finishes may be resistant to oil-borne or water-borne soil and stains or both. Stain and soil resistant finishes can be applied to fabrics used in clothing and furniture. Scotchgard is a stain and soil resistant finish commonly applied to carpet and furniture.

Oil and water proofing: These finishing allows no water to penetrate, but tend to be uncomfortable because they trap moisture next to the body. Recently, fabrics have been developed that are waterproof, yet are also breathable.

Water-repellent finishes: Water-repellent finishes resist wetting. If the fabric becomes very wet, water will eventually pass through. Applied to fabrics found in raincoats, all-weather coats, hats, capes, umbrellas and shower curtains.

Absorbent finishes: These finishing increase fibres moisture holding power. Such finishes have been applied to towels, cloth diapers, underwear, sports shirts and other items where moisture absorption is important.

Anti static finish: These finishing reduce static electricity which may accumulate on fibres. The most common type of anti-static finishes are fabric softeners.

Anti mildew: In certain ambient (humidity and heat) conditions, cellulose can be permanently damaged. This damage can be due to depolymerisation of the cellulose or to the fact that certain micro-oganisms (mildews) feed off it. The situation is worsened, during long storage periods, by the presence of starch finishing agents.This damage can be prevented by the use of antiseptics, bacteria controlling products containing quaternary ammonium salts and phenol derivatives. Dyestuffs containing heavy metals can also act as antiseptics. Permanent modification of the fibre (cyanoethylation) is another possibility.

Mothproofing finishes: These finishing protect protein-containing fibres, such as wool, from being attacked by moths, carpet beetles and other insects.

Antibacterial finish: The inherent properties of textile fibres provide room for the growth of micro-organisms. The structure and chemical process may induce the growth, but it is the humid and warm environment that aggravates the problem further. Antimicrobial finish is applied to textile materials with a view to protect the wearer and textile substrate itself.

Antimicrobial finish provides the various benefits of controlling the infestation by microbes, protect textiles from staining, discolouration and quality deterioration and prevents the odour formation.Anti-microbial agents can be applied to the textile substrates by exhaust, pad-dry-cure, coating, spray and foam techniques. The application of the finish is now extended to textiles used for outdoor, healthcare sector, sports and leisure.

UV protection: In UV protection the fabric is treated with UV absorbers which ensures that the clothes deflect the harmful ultraviolet rays of the sun, reducing a person's UVR exposure and protecting the skin from potential damage. The extent of skin protection required by different types of human skin depends on UV radiation intensity and distribution with reference to geographical location, time of day and season. This protection is expressed as SPF (Sun Protection Factor), higher the SPF value better is the protection against UV radiation.

Colour fastness improving finish: Colour fastness is the resistance of a material to change in any of its colour characteristics, to the transfer of its colourants to adjacent materials or both. Fading means that the colour changes and lightens. Bleeding is the transfer of colour to a secondary, accompanying fibre material. This is often expressed as soiling or staining meaning that the accompanying material gets soiled or stained. The physical and chemical principles involved in the performance of the fastness in improving finishes concern either the interaction with the dyestuff or with the fibre or both.

Plasma finish: Plasma treatment is a surface modifying process, where a gas (air, oxygen, nitrogen, argon,carbon dioxide and so on), injected inside a reactor which is ionised by the presence of two electrodes between which is a high-frequency electric field. The need to create the vacuum is justified by the necessity to obtain a so-called cold plasma with a temperature no higher than 80°C. This, with the same energy content that can be reached at atmospheric pressure at a temperature of some thousands of °C, permits the treatment of fabrics even with a low melting point such as polypropylene and polyethylene, without causing any form of damage. The fabric, sliding through the electrodes, is subject to a true bombardment from the elements that constitute the plasma (ions, electrons, UV radiation and so on) and which come from the decomposition of gas and contain a very high level of kinetic energy. The surface of the fabric exposed to the action of the plasma is modified, both physically (roughness), as well as chemically, to remove organic particles still present and to prepare for the successive introduction of free radicals and new chemical groups inside the molecular chain on the surface of the material. The mechanical properties remain, on the other hand, unaltered, as the treatment is limited to the first molecular layers.

7.4 Enzyme finishing in textile industry

Bio-polishing, also called bio-finishing, is a finishing process applied to cellulosic textiles that produces permanent effects by the use of enzymes.

Bio-finishing removes protruding fibres and slubs from fabrics, significantly reduces pilling, softens fabric and provides a smooth fabric appearance, especially for knitwear and as a pretreatment for printing.

7.4.1 Sewing thread finishing

Apart from many of the above said finishes which can be applied to sewing threads also. A variety of finishes are used to improve the sewability of sewing thread, for example:

1. Lubricants reduce friction and improve the lubricity of the thread. Lubricity refers to the frictional characteristics of thread as it passes through the sewing machine and into the seam. Good lubricity characteristics will minimise thread breakage and enhance sewability.

2. Glazing increases strength and abrasion resistance. Glaze finish refers to a finish put on 100% cotton threads or cotton-polyester core spun thread made from starches, waxes or other additives. This coating is then brushed to give the thread a smooth surface. A glaze finish protects the thread during sewing giving better ply security and abrasion resistance.

3. Bonding to increase strength and surface smoothness. Bonded finish refers to a finish applied to continuous filament nylon and polyester threads which coats the fibres, giving the thread better ply security and abrasion resistance.

7.5 Energy conservation in textile finishing processes

In textile finishing it is essential to reduce energy consumption and thermal pollution. This leads to an essential study on the ways of limiting energy wastage and thermal pollution in a textile finishing processes. The main activity involves singeing, desizing, scouring, hot washing, bleaching, mercerising, dyeing, printing, soaping, drying and finishing of cotton fabric. In this section, engineering measures to limit heat energy wastage discharged from continuous bleaching machine, drying chamber, boiler stack exhaust and blow down are discussed.

Textile finishing processes is one of the energy intensive industries all over the world. Tremendous energy waste is found in the textile finishing processes. With the heat recovery of discharged hot effluent and exhaust gases, not only the cost of production is reduced, but also the thermal pollution problems are solved without additional cost.

7.6 Identification of heat energy wastage in textile mill

After carrying out a detailed energy auditing on the production and non-production facilities, the cost-benefit analysis for the hot effluent from continuous bleaching machine, exhaust gas vapour from drying chamber and finishing processes, heat recovery from boiler stack and blow down are performed.

7.6.1 Hot effluent discharged from the bleaching machine

In textile industry the scouring and bleaching processes are integrated by a continuous bleaching machine. The two processes have an unique effluent. The temperature of hot drain wastewater and the flow rate of hot effluent for each machine are measured to 65°C and 7.3 m³/hr respectively. It is estimated that each continuous bleaching machine is in operation for 7000 hr per year. In order to achieve heat recovery from the hot effluent, there are two possible adaptable ways on the site.

One is to install individual heat exchanger for each of the three continuous bleaching machines. The hot effluent is directly fed into the heat exchanger, thus heating up the cold fresh water, which is then fed back to the continuous bleaching machine as shown in Fig. 7.1.

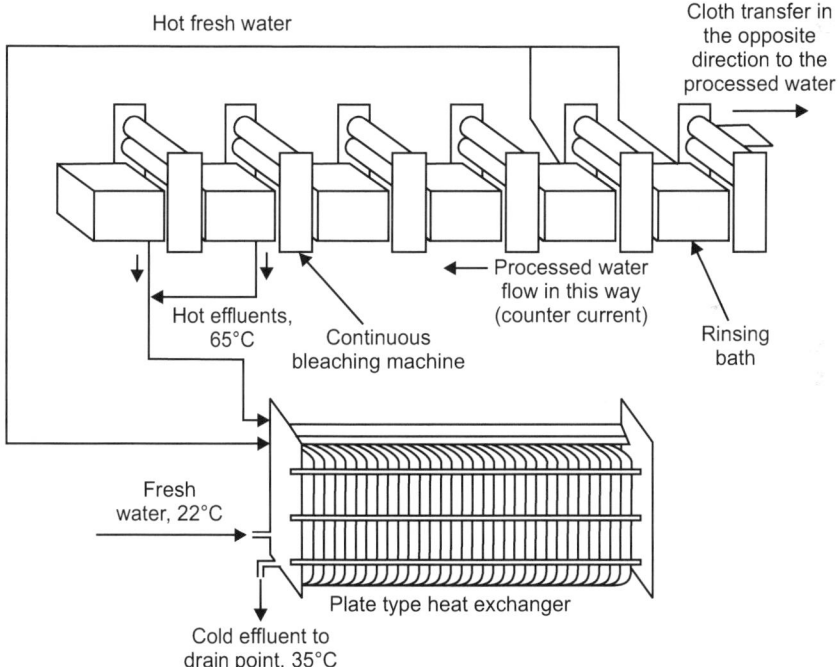

Figure 7.1: A heat recovery system for individual bleaching machine.

The second method is to collect all hot effluent from different machines into a collection sump. Then the waste heat can be recovered through a centralised system as shown in Fig. 7.2. After passing through the heat

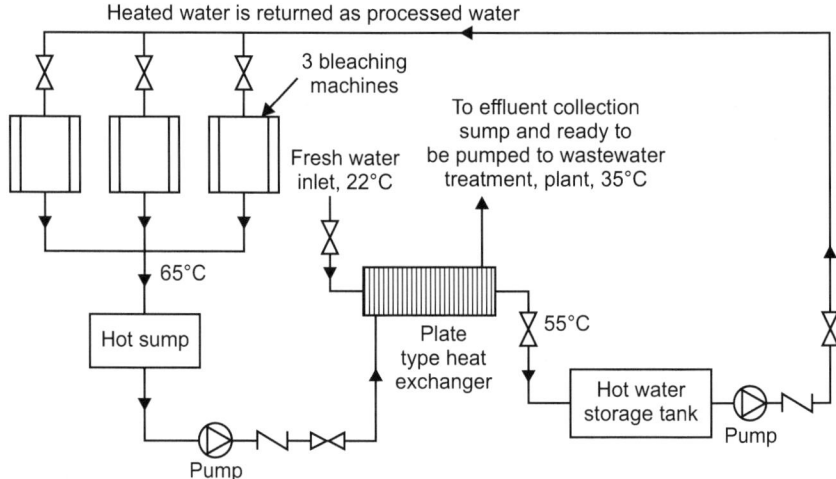

Figure 7.2: A centralised heat recovery system.

exchanger, the heated fresh water is pumped into a hot water storage tank. It is because sometimes not all the continuous bleaching machines are in operation simultaneously. The hot water is then distributed from the storage tank to the bleaching machines for use during production processes.

To decide on the best heat recovery method, following points should be considered according to the factory environment:

1. The continuous bleaching machines justify the extra cost raised by the centralised system. For the centralised heat recovery system, the costs of heat exchanger and pumping equipment, pipe works and control system are higher than installing three separate heat exchangers to three individual machines.

2. There is a large amount of heat loss in the collection sump tank and hot water storage tank for centralised system.

3. For individual heat recovery system, the heat exchanger operates only when the machine is in operation. This means that once it is in operation, the heated up fresh water can be used as processed water and fed back to the machine directly. The control is very simple and independent.

4. No pumping cost is required for the individual heat recovery because the hot effluent flow can be achieved by static head difference and the fresh water can be pumped by the water pressure inside the pipework.

It is found that the installation of a centralised heat recovery system is feasible and economically justified in the textile mill, if the number of continuous bleaching machine exceeds 10 and the total flow rate of hot effluent for heat recovery is more than 800 m^3/day. Hence an individual plate type heat exchanger for each of the three continuous bleaching machines are proposed.

In order to meet the effluent temperature requirement, the outlet temperature of hot drain water after passing through the heat exchanger is set to 35°C. The flow rate of specified fresh water, 22 °C is taken as 6.6 m^3/hr. The temperature of hot drain wastewater and the flow rate of hot effluent for each machine are measured to 65°C and 7.3 m^3/hr respectively. Hence the annual industrial fuel oil savings is estimated to 42,500 gallons (or 7200 × 10^6 kJ) for each plate type heat exchanger. This means an annual fuel cost savings of about $130,000.

7.6.2   Other energy conservation areas

Energy-efficiency improvements in stenters

Stenters have an important role in the dyeing and finishing of fabrics. Stenters are mainly used in textile finishing for heat-setting, drying, thermosol processes and finishing. It can be roughly estimated that, in fabric finishing, the fabric is treated on average 2–3 times in a stenter. A stenter essentially consists of a pair of endless traveling chains fitted with clips of fine pins and is carried on tracks. The cloth is firmly held at the selvages by the two chains which diverge as they move forward so that the cloth is brought to the desired width. Similar to heat setting and curing, stenters also affect the finished length, width and properties of the fabrics.

Fabric can be processed at speeds from 10–100 m/m and at temperatures of more than 200°C. Stenters can be heated in a variety of ways, such as direct gas firing and through the use of thermic fluid systems. Gas-fired stenters are highly controllable over a wide range of process temperatures.

Thermic fluid heating for stenters requires a small thermic oil boiler (usually gas-fired) and its associated distribution pipeline. This system is less efficient than direct gas firing and has higher capital and running costs. However, like gas, it can be used over a wide temperature range, but the problem is that this heating can only be done indirectly via a heat exchanger. This system, compared with indirect gas firing, is relatively inefficient, so it is no longer commonly used. Finally, there are a number of steam-heated stenters. Because of their low temperature limits (usually up to a maximum of 160°C) these stenters can only be used for drying; they are not suitable for heat setting or thermo fixing of fabrics. In all stenters the hot air is blown against the fabric and then recirculated. A fraction of this air is exhausted and made up with fresh air. To provide better control, stenters are split-up into a number of compartments, usually between two and eight.

A typical energy breakdown for a stenter being used for hot-air drying is shown in Table 7.1. By far the greatest users of energy are the evaporation and air heating components. It is therefore necessary that the fabric moisture content is minimised before the fabric enters the stenter and that exhaust airflow within the stenter is reduced. Many stenters are still poorly controlled, relying on the manual adjustment of exhausts and operator estimations of fabric dryness.

Table 7.1: Energy breakdown for a typical stenter.

Component	Energy use (GJ/T of product)	Share of energy use from total energy use
Evaporation	2.54	41.0 %
Air heating	2.46	39.7 %
Fabric	0.29	4.6 %
Case	0.39	6.3 %
Chain	0.09	1.5 %
Drives	0.43	6.9 %
Total	6.20	100 %

Conversion of thermic fluid heating systems to direct gas firing systems in stenters and dryers

Often, thermic fluid heaters are used to provide the heating requirements of stenters and dryers. In this system, a fluid is heated up to 260°C and circulated in the plant through transmission lines. Heat is transferred from the hot fluid to the chambers using radiators. Substantial heat loss happens in the thermic fluid boilers, transmission lines and radiators. To reduce this heat loss, thermic fluid heating systems can be replaced with direct gas firing systems.

The direct gas firing system has several advantages over the thermic fluid heating system. First, there is a saving on fuel consumption with the reduced heat losses. The electricity required for pumping the thermic fluid and the risks involved in the circulation of the hot fluid are also eliminated. Besides, the direct gas firing system has the advantage of direct heating at the chamber itself. The temperature in the processes can be attained as per process requirements, without time lag. There is no dependency on the centralised heating system and better production planning can be achieved for the process. Heating up times are shorter in stenters with direct gas firing systems, which leads to reduced idling losses. When direct gas firing systems were first introduced there was concern that oxides of nitrogen, formed by the exposure of air to combustion chamber temperatures, would cause fabric yellowing or partial bleaching of dyes. This fear has been shown to be unjustified.

New machines have successfully been designed incorporating direct gas firing systems. In the modified design, the physical size and shape of the drier have not been altered and the existing air passage has been converted into a combustion chamber. A textile plant converted their thermic fluid system into a direct gas firing system can achieve 11000 GJ/year savings in fuel use (around 40% of the total fuel use) and 120 MWh/year savings in electricity use (around 90% of the total electricity use). The total investment cost of this retrofit measure was around US$50,000 with a payback period of one year.

Introduce mechanical de-watering or contact drying before stenter

Stentering is an energy intensive process, so it is important to remove as much water as possible before the fabric enters the oven. This can be achieved using mechanical dewatering equipment or by using contact drying using heated cylinders. Up to 15% energy savings in the stenter (depending on the type of substrate) can be obtained if the moisture content of the fabric is reduced from 60% to 50% before it enters the stenter. Contact drying is roughly five times more energy intensive than suction slot de-watering, but uses only half to two-thirds the energy of a stenter. Pre-drying fabric to about 25–30% moisture content before passing it through the stenter still allows the fabric width to be adjusted to suit customer requirements. Other techniques used to reduce drying costs include infra-red and radio frequency drying. Gas-fired infra-red has been used for the pre-drying of textiles prior to stentering. This can have the effect of increasing drying speed by up to 50%, thereby relieving production bottlenecks which tend to occur at stenters.

In addition, energy savings in the range of 50–70% can be achieved compared to conventional stenter drying. Radio frequency drying is used extensively for the drying and dye fixing of loose stock, packages, tops and hanks of wool and sewing cotton. The energy requirement of radio frequency drying is approximately 70% of that of a conventional steam-heated dryer. However, its use is limited to loose stock and packages. It cannot be modified, as yet, to accommodate knitted or woven fabric since the traditional pins and clips of the Stenter transport mechanism interfere with the radio frequency drying field, causing discharge.

Avoid overdrying in stenter

The high-energy cost of running a stenter means that it is vital to avoid over-drying. Automatic infra-red, radioactive or conductivity-based moisture measurement systems can be linked to the stenter speed control to ensure that the appropriate fabric moisture content is achieved.

Close exhaust streams during idling in stenter

It is common practice to leave the exhaust systems running during the changing of batches, which may take 10–15 min or more and in some extreme cases (in some commission dyers) could happen every hour. Since stenters have a large air-heating requirement, it is important, whenever possible, to isolate the exhausts, or at least partially close them during idle periods. Proper scheduling in finishing minimises machine stops and heating-up/cooling-down steps and is therefore a prerequisite for energy savings.

Drying at higher temperatures in stenter

Drying at a higher temperature, if the fabric will tolerate it, means that radiation and convection losses become relatively small compared to evaporation energy, thereby reducing the total energy use per unit of product.

Close and seal side panels in stenter

On older machines, side panels may become damaged, distorting the air balance within the oven sections. All faulty panels should be repaired or replaced to provide an effective seal around the oven.

Proper insulation

Proper insulation of stenter envelopes reduces heat losses to a considerable extent. Improving stenter insulation is not usually practicable, although on some older machines it may be cost-effective to insulate the roof panels. Savings in energy consumption of 20% can be achieved if the insulation thickness is increased from 120 to 150 mm (provided that the same insulation material is used).

Optimise exhaust humidity in stenter

Table 7.1 shows that the main energy requirements for a stenter are for air heating and evaporation. In order to optimise drying rates and energy use, air flows through the oven (and therefore the exhaust rate) must be carefully controlled. A significant number of stenters still rely on manual control of exhausts, although this is actually very difficult and often means that exhausts are left fully open unnecessarily. Energy consumption for air heating can reach up to 60% of the total energy requirement (compared to the 39.7% as shown in Table 7.1) if airflow is not monitored. For optimum performance, exhaust humidity should be maintained between 0.1 and 0.15 kg water/kg dry air. It is not unusual to find stenters with an exhaust humidity of only 0.05 kg water/kg dry air, indicating that the exhaust volume is too great and excessive energy is being used to heat air.

Equipment is available which will automatically control dampers to maintain exhaust humidity within a specified range, thereby reducing air losses without significantly affecting fabric throughput. However, there are some fabrics and processes (notably pre-setting of some synthetics) which cause considerable fume problems, leading to a 'blue haze' emanating from the fabric slots. In these cases the stenter may be required to operate with fully open exhausts. Another type of equipment is variable speed fans, which will automatically adjust exhaust airflow according to the moisture content of the exhaust air or according to the moisture content or temperature of the fabric after the process. A reduction of fresh air consumption from 10 kg fresh air/kg textile to 5 kg fresh air/kg textile results in 57% energy savings. A case-study in a synthetic textile plant in India intended to control the exhaust air fan as per requirement and reduce losses by using a semi-automatic control system reported fuel savings of 670 GJ/year with an investment cost of only US$600.

Install heat recovery equipment in stenter

There are two types of heat recovery equipment that can be used to recover heat from stenters' exhaust gases. These are:

1. Heat-recovery air/air: Uses exhaust air heat to heat up fresh air supplied to the stenter.
2. Heat-recovery air/water: Uses exhaust air heat to heat up service water for wet finishing (for example washing, dyeing and bleaching.)

Exhaust heat recovery can be achieved by using air-to-air systems such as plate heat exchangers, glass tube heat exchangers or heat wheels. Efficiencies are generally about 50–60%, but there can be problems with air bypass, fouling and corrosion.

Air-to-water systems, such as spray recuperation, avoid fouling and clean the exhaust, but may increase corrosion. Secondary water/water heat exchange equipment is required and a matching heat requirement must be identified. The resultant hot water can be used in dyeing. Approximately 30% savings in energy can be achieved. If large quantities of volatile organics or formaldehyde are generated by the stenter, some form of scrubber, electrostatic precipitator or even an incinerator may be required to comply with limits set by relevant environmental regulations. Due to the cooling of the outgoing airflow, pollutants that are in the air are condensed and can be removed with a filter. In this way organic pollutants are eliminated from the air emissions and the recovered heat can be reused. The investment cost varies from US$77,000 to US$460,000.

Minimising energy consumption in stenters, especially if heat recovery systems are installed, requires adequate maintenance (cleaning of the heat exchanger and stenter machinery, checking of control/monitoring devices, etc.).

The above information does not consider the installation of other measures such as fabric moisture controls and exhaust humidity controls. If these systems are installed, according to some sources heat recovery may not be cost-effective.

Efficient burner technology in direct gas-fired systems in stenter

Optimised firing systems and sufficient maintenance of burners in direct gas-fired stenters can minimise methane emissions, which is important because methane emissions from burners greatly determine actual burner capacity. Stenters should receive general maintenance by specialised companies at regular intervals. There should also be routine checking of the burner air inlet for blockings by lint or oil, cleaning of pipe works to remove precipitates and adjusting of burners by specialists.

Use of sensors and control systems in stenters

Sensors and control systems are very important for stenters in order to assure quality control as well as the efficient use of energy. Some of the major control systems that are used in stenters are discussed below:

1. Exhaust humidity measurement: Measuring and controlling humidity to load exhaust air most efficiently with humidity. This reduces the volume of hot exhaust air and thus energy waste is dramatically.
2. Residual moisture measurement: Residual moisture control provides the highest productivity at the lowest energy cost. Overdrying and overheating of fabric can be avoided by using this control system.
3. Fabric and air temperature measurement: Several fabric temperature sensors placed inside the stenter along with the fabric provide a good system for the supervision and optimisation of heat treatment processes.
4. Process visualisation systems: These systems show the process parameters and how different parts of the machine are performing in real time and gives the operator useful information in order to better monitor the performance of the machine.

As per latest findings, the capital cost for moisture humidity controllers for dryers and stenters to be in the wide range of US$20,000–220,000 with a payback period between of 1.5 and 5 years. Also, it has been reported that the capital cost for the dwell time controls system for dryers and stenters, which is one of the important control systems that can be installed in these machines, in the wide range of US$80,000–400,000 with a payback period of between 4 and 6.7 years. Figure 7.3 shows the review of major energy-efficiency opportunities in stenter machines.

Some good housekeeping measures for stenters are:

1. It is recommended to utilise at least 75–80% of the width of the Stenter.
2. Periodic cleaning of filters is necessary since clogged filters will impair drying efficiency,
3. Ensure leak proof chamber doors and the adequate insulation of the top, bottom and sides.

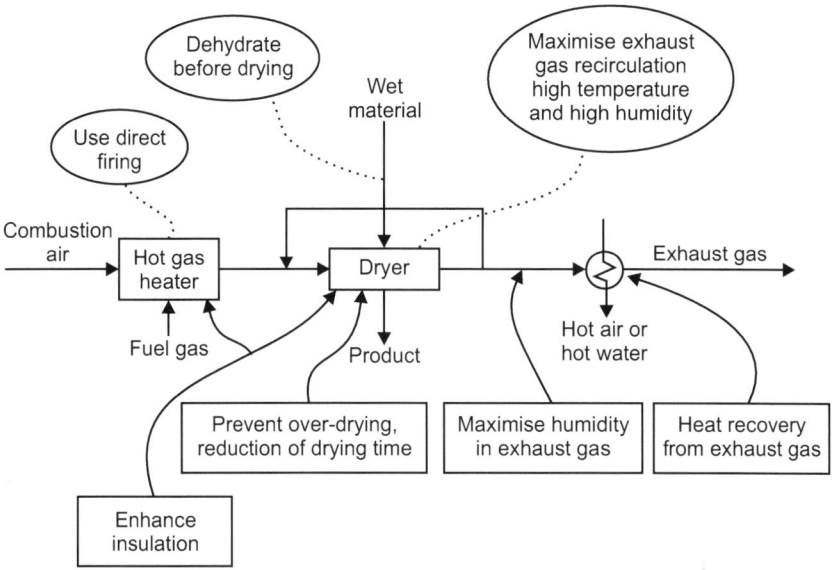

Figure 7.3: Energy-efficiency opportunities in stenter machines.

4. The blower motor should be interlocked with the main drive, i.e., it will stop when the machine stops.

7.6.3   Heat recovery from drying chamber

For the drying process, the cloth enters a drying chamber after rinsing. Then it is rolled through a drying drum which is heated up by steam from boiler as shown in Fig. 7.4. Steam is drawn into the drying drum to heat up the drum surface and then the cloth rolling through the surface is heated up and the water content will be evaporated, making the cloth dry.

The condensation inside the drum will then be discharged through a pipe by the pressure built up in the drum from the steam. This is different from the other drying method in which the water vapour is drawn by keeping a low humidity in the drying chamber.

In considering the design of a heat recovery system for the drying chamber, there is a choice between using the water vapour to heat up water or to heat up air. From the prescribed drying principle in the factory, there is only one possible way by which the heat recovered can be used to heat up the water and then to recycle this heated water to the rinsing batch as processed water. It is not suitable to heat up the air because there is difficulty in utilising this hot air in most factories. However, in places with cold weather, the hot air can be utilised as space heating for buildings.

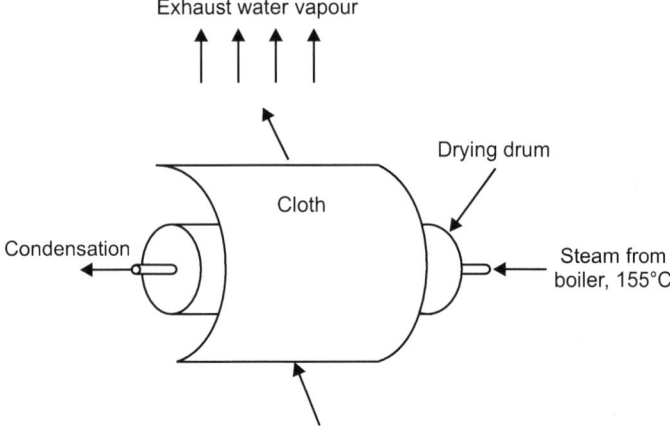

Figure 7.4: An inside view of a drying chamber.

The temperature and total flow rate of exhaust water vapour are measured to 52°C and 70,000 m³/hr respectively. The heat recovery from the dryer exhaust is feasible theoretically. Hence the annual industrial fuel oil savings is estimated to 40,500 gallons (or 6,850 × 10⁶ kJ). This means an annual fuel cost savings of about $ 124,000. However, having reviewed the potential energy savings from installing a water-heating economiser under the condition of 52°C exhaust steam, the energy savings is too low to justify the cost of an economiser.

7.7 Heat recovery from boiler stack exhaust and blow down

Boiler exhaust gas rejected from the stack is normally of 40–50°C higher than the temperature of generated steam. By recovering this heat, the consumption of fuel oil and also the thermal pollution to the environment can be reduced. Heat can be recovered by either the economiser to heat feed water or by air pre-heater for combustion. However mills are already been incorporated with air pre-heaters and other auxillaries.

It is worth noted that the exhaust gas temperature can be reduced only above the dew point of flue gas, otherwise corrosion problem may occur due to the formation of sulphuric acid. The supplier of carbon steel economiser recommends that the temperature of the feedwater has to be near 104°C if the dew point is about 128°C for a sulphur content of 0.5 % in order to prevent corrosion. Thus, a pre-heat skid is developed for heat economiser feedwater by using boiler steam as shown in Fig. 7.5. Because the steam energy used in the pre-heat skid is recycled back into the boiler, there is no significant net energy used to pre-heat the feedwater. Then, economisers can be installed as shown in Fig. 7.6 on the site to suit boilers by connecting a lower transition.

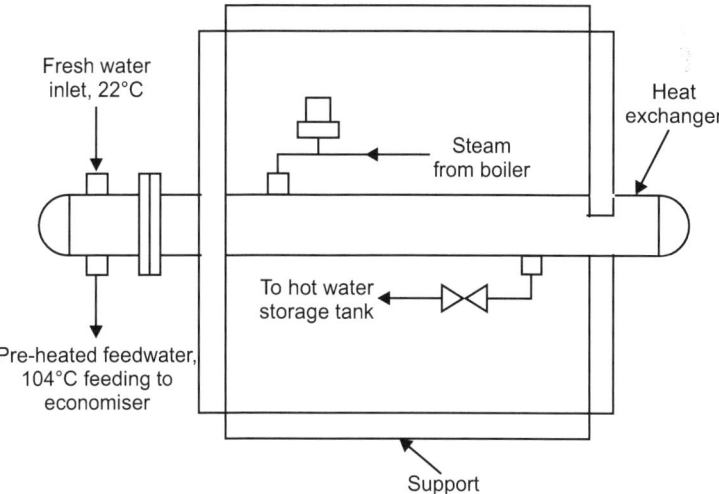

Figure 7.5: A schematic diagram of a feedwater pre-heated skid.

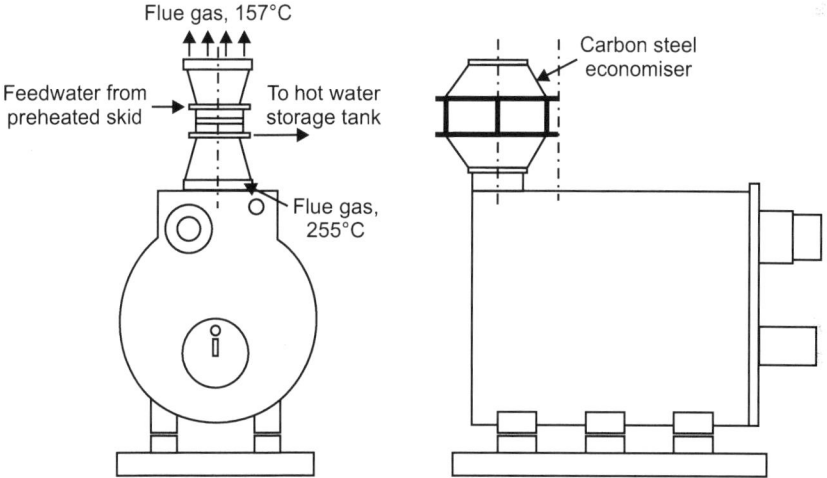

Figure 7.6: A schematic diagram of an economiser in a boiler.

The exhaust from boilers has no economical benefit in the recovery of heat. The discrepancy flow rate between these two boilers is due to the different operating conditions of the boilers. Thus, the outlet temperature of the flue gas is 157°C and the inlet temperature of feedwater from the pre-heated skid is 104°C. The boiler is in operation for about 7800 hr per year. Hence the annual industrial fuel oil savings is estimated to 107,000 gallons (or 18,000 × 10^6 kJ) for installing the economiser. This means an annual fuel cost savings of about $ 330,000.

7.8 Heat recovery from boiler stack blow down

The process of boiler blow down is to maintain a low concentration of dissolved and suspended solids in the boiler water and to remove sludge in the boiler in order to avoid priming and carryover.

The blow down recovery system is designed to recover both the latent heat of the 156°C steam in the first closed type shell and tube heat exchanger and sensible heat of the 100°C water in the second plate heat exchanger as shown in Fig. 7.7. The blow down is forced to pass through the first heat exchanger by the high pressure inside the boiler drum and then effluent contained suspended solids will be discharged at the second plate type heat exchanger at a temperature of 40°C. Afterwards, the hot feed water is pumped into the hot water storage tank and ready to feed back to boiler. In fact, all heated effluents including those from the economiser and steam condensation in the plant will also return to the hot water storage tank. If overflow occurs, the feedwater will flow to a temporary storage tank.

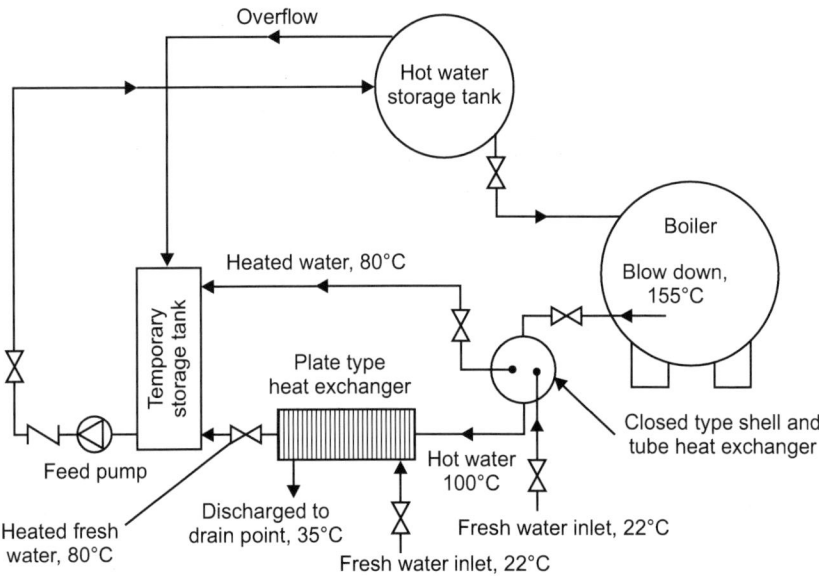

Figure 7.7: A schematic diagram of a blow down heat recovery system.

With the existing boiler capacity of 4.4 kg/s at 5.5 bar pressure and the blow down flow of about 4 %, hence the annual industrial fuel oil savings is estimated to 17,900 gallons (or 3000 × 10⁶ kJ) from boiler blow down. This means an annual fuel cost savings of about $55,000.

Energy conservation in man-made fibre industry

8.1   Introduction

Man-made fibres are spun and woven into a huge number of consumer and industrial products, including garments such as shirts, scarves and hosiery; home furnishings such as upholstery, carpets and drapes and industrial parts such as tyre cord, flame-proof linings and drive belts. The chemical compounds from which man-made fibres are produced are known as polymers, a class of compounds characterised by long, chainlike molecules of great size and molecular weight. Many of the polymers that constitute man-made fibres are the same as or similar to compounds that make up plastics, rubbers, adhesives and surface coatings. Indeed, polymers such as regenerated cellulose, poly-caprolactam and polyethylene terephthalate, which have become familiar household materials under the trade names rayon, nylon and Dacron (trademark), respectively, are also made into numerous nonfibre products, ranging from cellophane envelope windows to clear plastic soft-drink bottles. As fibres, these materials are prized for their strength, toughness, resistance to heat and mildew and ability to hold a pressed form.

Man-made fibres are to be distinguished from natural fibres such as silk, cotton and wool. Natural fibres also consist of polymers (in this case, biologically produced compounds such as cellulose and protein), but they emerge from the textile manufacturing process in a relatively unaltered state. Some man-made fibres, too, are derived from naturally occurring polymers. For instance, rayon and acetate, two of the first man-made fibres ever to be produced, are made of the same cellulose polymers that make up cotton, hemp, flax and the structural fibres of wood. In the case of rayon and acetate, however, the cellulose is acquired in a radically altered state (usually from wood-pulp operations) and is further modified in order to be regenerated into practical cellulose-based fibres. Rayon and acetate therefore belong to a group of man-made fibres known as regenerated fibres.

Man-made fibres can be divided in two main categories: artificial fibres and synthetic fibres. Artificial fibres are derived from natural products (in most cases cellulose) that are modified by reactive agents.

Most known artificial fibres from cellulose are given below:

1. Rayon or viscose.
2. Modal.

3. Lyocell.
4. Cuprammonium.
5. Acetate.
6. Triacetate.

Another group of man-made fibres (and by far the larger group) is the synthetic fibres. Synthetic fibres are made of polymers that do not occur naturally but instead are produced entirely in the chemical plant or laboratory, almost always from by-products of petroleum or natural gas. These polymers include nylon and polyethylene terephthalate, mentioned above, but they also include many other compounds such as the acrylics, the polyurethanes and polypropylene. Synthetic fibres can be mass-produced to almost any set of required properties. Synthetic fibres have been commonly used in the textile industry for several years both for apparel and for furnishing fabrics and in recent years they have evolved into new technical fibres.

Synthetic fibres are derived from substances that are not present in nature, but instead created through chemical reactions (synthesis) from petrochemical products. Both types of fibres can be produced as a filament or can be cut at predifined lengths. When the fibre length is between 30 mm and 60 mm, the fibres are considered short or commonly called 'cotton cut' and can be processed in standard spinning mills. Synthetic fibres have been commonly used in the textile industry for several years both for apparel and for furnishing fabrics and in recent years they have evolved into new technical fibres.

In recent years textile products for technical applications in various high-tech areas have experienced great advancements. These materials are composed of natural, artificial and synthetic fibres and are capable of offering special characteristics such as mechanical, thermal conduction, durability, resistance to flame, heat, smoke, chemicals, etc. Designed and produced based on particular needs, they are capable of performance.

Processing of technical fibres: The transformation process from fibre to yarn to fabric is comparable to the traditional process used for cotton fibres: opening, spinning, weaving, finishing, etc.

Spinning process: Although technical and synthetic fibres do not require cleaning and are of uniform length, the spinning process is not all easy or certain; rather, the choice of equipment and technology, as well as production and spinning planning, need to be appraised with care. Only with the best knowledge of the fibres to be employed one could select the right equipment to obtain the production and the quality to compete on an international level. The spinning process must be adaptable to the raw materials used in order to produce yarns that better showcase the fibre's technical characteristics.

Opening: One of the challenges of working with technical fibres is the opening of the fibrous mass without scraping, breaking, weakening the fibre. It is therefore of utmost importance to choose the correct composition of opening and blending equipment that will maximise production and quality.

Blending: When processing synthetic and technical fibres, different colours of the same fibre or fibres of different composition are often blended together.

Carding: Carding is a mechanical process that disentangles, cleans and intermixes fibres to produce a continuous web or sliver suitable for subsequent processing. This is achieved by passing the fibres between differentially moving surfaces covered with card clothing. It breaks up locks and unorganised clumps of fibre and then aligns the individual fibres to be parallel with each other.

8.2   Manufacturing process of acetate rayon

Raw materials: Raw materials used in manufacture of acetate rayon are – purified cotton linters, wood pulp, acetic anhydride, acetic acid and sulphuric acid.

The following steps are involved in manufacturing of acetate rayon.

Step 1: Activation with acetic acid

The process involves steeping of purified cotton in acetic acid which makes swelling and makes cellulose more reactive.

Step 2: Acetylation

The pre-treated cotton with acetic acid is then acetylated with excess acetic acid, acetic anhydride, with sulphuric acid to promote the reaction.

Step 3: Hydrolysis of triacetate

The triacetate formed is hydrolysed to convert triacetate to diacetate. The resultant mixture is poured in water to precipitate the cellulose acetate.

Step 4: Spinning

Secondary acetate is dissolved in acetone, filtered, dearerated and passed through spinerrate in hot air environment, which evaporates the solvent.

Economical production:

1. Low cost and availability of acetic acid and acetic anhydride.
2. Recovery of acetic acid.
3. Recovery of acetone.

Process flow diagram for the manufacture of acetate rayon is given in Fig. 8.1.

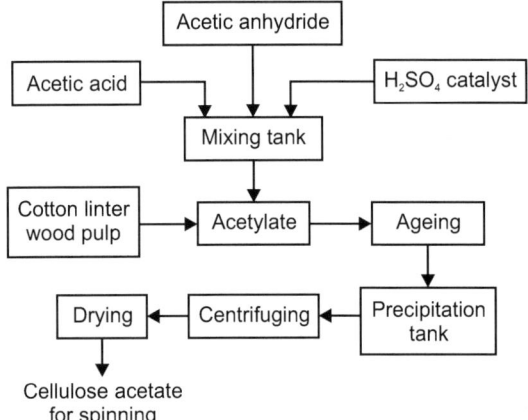

Figure 8.1: Process flow diagram for the manufacture of acetate rayon.

8.2.1 Cuprammonium rayon

Cuprammonium rayon is made from reaction of cellulose with copper salt and ammonia. After bleaching cellulose is added in ammonical solution of copper sulphate resulting in formation of cuprammonium cellulose which is spun into water and the yarn is washed with acid to remove traces of ammonia and dried. Process flow diagram for the manufacture of cuprammoium rayon is given in Fig. 8.2.

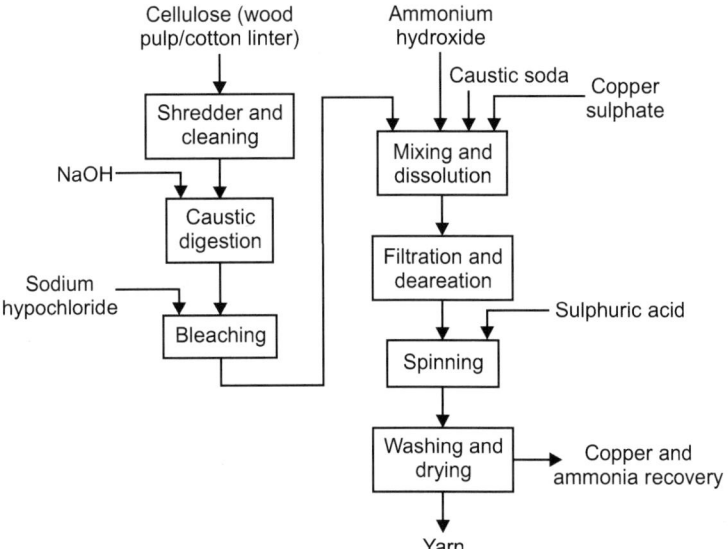

Figure 8.2: Process flow diagram for the manufacture of cuprammonium rayon.

8.3 Energy-efficiency technologies and measures in man-made fibre production

8.3.1 Installation of variable frequency drives (VFD) on hot air fans in after treatment dryers in viscose filament production

The raw material for cellulose fibre production (rayon) is wood pulp which is an inexpensive and renewable resource, however the processing of the wood pulp into rayon is a highly energy and water intensive process. The production of rayon involves three stages: (i) dissolution of the wood pulp, (ii) extrusion of the yarn and (iii) purification of the yarn by bleaching and washing. The purified rayon is dried using hot air driers.

The drying process of the yarn cakes is the final stage of rayon production and it is carried out using air dryers which circulate hot air to remove moisture from yarn cakes. The drying is carried out in various stages in different zones of the dryer. Maximum drying occurs in the first two zones and drying gradually reduces in subsequent zones. In the last zone, the moisture removal rate is very low and is independent of air velocity. Hence the speed of air circulation fans in the last zone, where the drying rate is independent of air velocity, can be reduced. In one of its plants, Century Rayon Company, the largest viscose filament yarn producer in India, modified four of its after treatment dryers in order to optimise the air velocity in the last zones of the dryer. The new system consists of the installation of variable frequency drives (VFD) for a group of motor driven hot air fans and these fans are grouped based on the zones. In the initial stage where the amount of moisture is high, the VFDs are operated at higher speeds (rpms) to deliver more hot air flow and thus reduce the moisture content. As the yarn passes through the second stage the speed of the motors is reduced as the moisture removal is reduced from that of the initial stage. In the final stage the amount of moisture is very minimal and thus the speed of the hot air fans is further reduced. Fan speed reduction using a variable frequency drive for a group of motors has resulted in significant energy savings.

8.3.2 Use of light weight carbon reinforced spinning pots in place of steel reinforced pots

Conventionally, steel reinforced spinning pots are used in synthetic fibre production plants. In Century Rayon Company's plant in India, the weight of a steel reinforced spinning pot was 2.8 kg and the energy consumption per spinning machine per day was about 581 kWh. Steel reinforced spinning pots can be replaced with carbon reinforced spinning pots. The weight of the carbon reinforced pots used in Century Rayon Company's plant is 2.2 kg (Fig. 8.3).

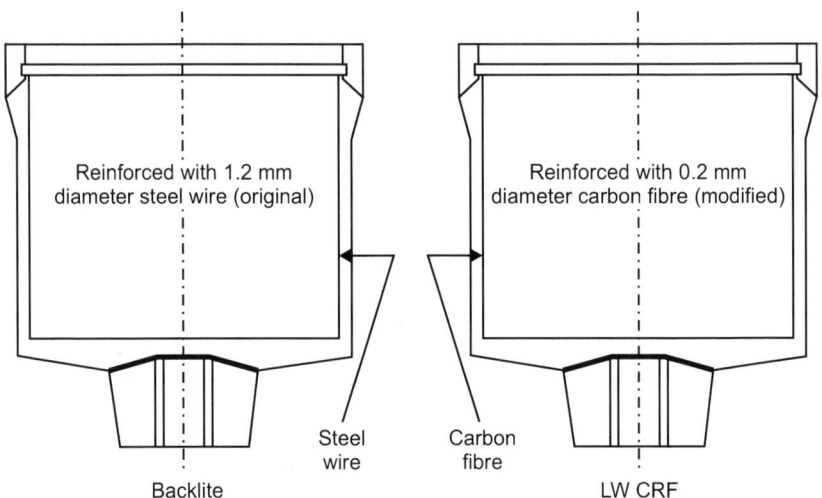

Reinforced with 1.2 mm diameter steel wire (original)

Reinforced with 0.2 mm diameter carbon fibre (modified)

Steel wire

Carbon fibre

Backlite

LW CRF

Figure 8.3: Steel reinforced and carbon reinforced spinning pots.

After the replacement, the energy consumption per spinning machine per day was reduced to 549 kWh. The reduction in weight of spinning pots thus resulted in a reduction in energy consumption of 9.6 MWh per spinning machine per year. The investment cost of the replacement was around US$680 per machine. Century Rayon Company implemented this retrofit measure on 39 spinning machines and reported a payback period of 9 months for the investment.

8.3.3 Installation of variable frequency drives in fresh air fans of humidification systems in man-made fibre spinning plants

In some parts of the production process of man-made fibres, the temperature and humidity of the ambient air should be controlled and maintained according to certain requirements. Therefore, in man-made fibre production plants, there are several air conditioning and humidification systems. Variable frequency drives can be installed on fresh air fans of the humidification system.

The speed of all these fans can be controlled by changing the frequency of the VFD as to meet the requirement for fresh air by measuring the temperature as well as the relative humidity (RH) of the ambient air in the production process. A viscose filament production plant in India installed VFDs on 19 fresh air fans in its humidification systems. This resulted in an average electricity savings of 32.8 MWh/fan/year. The investment cost is reported to be around US$5600 per fan.

8.3.4 Installation of variable frequency drives on motors of dissolvers

Dissolvers are used in the production of viscose filaments, for example in the immersion step where the cellulose is dissolved in caustic soda. The dissolvers normally run with a fixed speed, while the speed could be adjusted based on the process requirements. A viscose filament production plant in India installed VFDs on six dissolvers for speed control as per the process requirements. They have reported an average electricity savings of 30.3 MWh/dissolver/year. The investment cost was around US$16,400 per dissolver.

In another plant belonging to Century Rayon Company, every batch in the dissolver takes 180 min, during which the agitator was run at a fixed speed of 1000 rpm. The company decided to run the agitator at lower speed of 500 rpm for the last 60 min (1000 rpm for the first 120 min) by installing VFDs. This resulted in electricity savings of 49.5 MWh/agitator/year with an investment cost of about US$9500 per agitator. There was no impact observed on the viscose quality.

8.3.5 Adoption of pressure control systems with VFDs on washing pumps in the after treatment process

Variable frequency drives with pressure controls can be installed on washing pumps in the after treatment process of viscose yarn production. In this way, the energy consumption of washing pumps can be optimised by varying the speed of the pumps as per the variation in the denier (thickness) of yarn. A viscose filament production plant in India installed this measure on 19 washing pumps and on average achieved an electricity savings of 40.4 MWh/pump/year with an investment cost of US$930 per pump.

8.3.6 Installation of lead compartment plates between pots of spinning machines

In spinning machines, spinning pots are rotating at around 7800 rpm. This produces cross currents of air between adjacent pots. This cross current of air increases the electrical load of machines. Lead compartment plates can be installed between each spinning pot to overcome the cross current of air between the pots. Century Rayon Company in India installed lead compartment plates in 69 spinning machines.

They have reported an average electricity savings of 7 MWh/machine/year. The investment cost of the installation is not reported, but it is said that the payback period is immediate.

8.3.7 Energy-efficient high pressure steam-based vacuum ejectors in place of low pressure steam-based vacuum ejectors for viscose deaeration

The function of viscose deaerator is to remove trapped air bubbles from filtered viscose before it is pumped to spinning machine for regeneration. Conventionally, a low pressure, steam-based four stage vacuum ejector is used for this purpose. However, low pressure steam-based four stage ejectors are sometimes not able to create the required vacuum due to less steam pressure. This results in higher steam consumption. To solve this problem, two stage high pressure steam-based ejectors can be installed in place of low pressure steam-based ejectors. The implementation of this measure in a plant in India resulted in fuel savings of about 3800 GJ/year with an investment cost of US$29,000.

8.3.8 Use of heat exchangers in dryers in viscose filament production plants

In dryers in viscose filament production lines, fresh air is usually heated through steam coils and then is circulated to the drying chambers. The moist vapour is then exhausted to the atmosphere at around 90°C. A heat exchanger can be installed in order to recover the heat in the exhaust air and use it to preheat the fresh air. This will result in significant fuel savings.

The energy savings from the installation of heat exchangers can be as high as 1 GJ/hr of dryer operation. The investment cost for the heat exchanger is reported to be around US$66,700.

8.3.9 Optimisation of balloon setting in TFO machines

Research has shown that TFO (two-for-one) twister machines consume less electricity at lower yarn balloon settings. However, the balloon size cannot be simply just lowered; rather the balloon setting should be optimised by conducting various studies with respect to different yarn counts and yarn twists. An electricity savings of 205 MWh/year is reported as a result of implementation of this measure, but the number of machines on which the measure was implemented was not specified.

8.3.10 Solution spinning high-speed yarn manufacturing equipment (for filament other than urethane polymer)

This equipment achieves high-performance and energy savings while employing new technology for the manufacture of yarn from raw materials such as rayon. The equipment is comprised primarily of the raw material mixing equipment and the spinning head. The polymer is extruded by a gear pump through the spinning nozzle while being dissolved, filtered and degassed in the raw liquid mixing equipment. In the spinning head, the raw liquid exiting the nozzle is passed through the primary, secondary and tertiary spindle units for molecular alignment and fully coagulated to form the yarn.

In conventional equipment, an efficient electric motor providing high-speed spinning has been adopted to permit. The high-speed and energy-efficient motors employed have increased spinning speed by a factor of approximately 2.7 (300 m³/min → 800m³/min) and also reduced electricity use by about 35%. Electricity savings can be as high as 500 MWh/machine (16 spindles)/year. Furthermore, this equipment has few consumable components, dramatically reducing maintenance costs. The investment cost for the installation of this measure (including construction cost) is about US$200,000. There are many examples of installation of this equipment in large plants throughout Japan.

8.3.11   High-speed multiple thread-line yarn manufacturing equipment for producing nylon and polyester filament

This equipment is different from the one explained above and it melts and spins nylon and polyester filament to spin fully drawn filament (FDY) or partially oriented filament (POY) at a high speed of 6000m/min. In addition, drawn winding is performed on the same process to enhance energy efficiency and productivity. Conventional equipment types are batch or indirect continuous, because spun yarn is produced in the spinning and then the final filament (FOY or POY filaments) is produced in drawing machines. The new equipment performs the spinning process and filament making process continuously, resulting in a 55% decrease in electricity consumption. The investment cost for the installation of this equipment is about US$320,000.

Recovery of reuse of chemicals in textile industry

9.1   Introduction

Chemical recovery system is an integral part of textile industry and the efficiency of chemical recovery plays and important role in economics. The continued increase in the cost of chemicals, energy and water makes their recovery more important today than it was 35 years ago when ultrafiltration and hyper filtration were first introduced to the textile industry. While the filtration techniques have been used at only a few installations, these plants have been able to save enough to pay for the recovery process in one to two years.

One key to having a successful recovery operation is to have good automatic control of the process.

This can drastically improve the economics of the textile process as well as minimising the cost of the recovery system.

The early application of ultrafiltration to PVA size recovery was accepted and used in a full plant scale installation almost 35 years ago. The process was successful then and continues to perform well. One limitation of the use of ultrafiltration was diversity of sizing chemicals that are in use. Controlling the size used on greige fabric is also necessary if the plant wants to recover PVA size from its preparation range. Inspite of these limitations, there has been an increase in the number of plants that are recovering PVA size finish. While recovery of size is only done in a small fraction of the finishing plants, interest in the recovery process is strong and the need appears to be increasing. As hazardous waste regulations increase and restrictions on the discharge of trace metals increase, the textile industry will be faced with the continual upgrading of its treatment facilities. The increase in the cost of solid waste disposal and the effect this will have on sludge disposal. While this is not directly connected to the recycle process, it illustrates the continued increases in the cost of all aspects of waste treatment.

In addition to the cost of solid waste disposal, the cost of energy is much higher today than it was almost 35 years ago when the recovery of PVA was first introduced. Since all of these factors continue to place pressure on the industry to eliminate waste discharges, several processes where recycling may be applied or is being applied with considerable success is discussed.

9.2 PVA recovery

Recovery of Polyvinyl Alcohol (PVA) was one of the first recycle processes to be used by the textile industry. A diagram of a PVA recovery system is shown in Fig. 9.1. An important factor is that when size is recovered, 60,000 gallons of discharge from each range per day is eliminated from the waste stream. While some blow down or clean up wastes are still discharged, the volumes are small and only contribute a fraction to the unrecycled waste stream. Costs of PVA, fresh water and waste treatment have increased significantly. Recycling of PVA is more attractive today than it was previously.

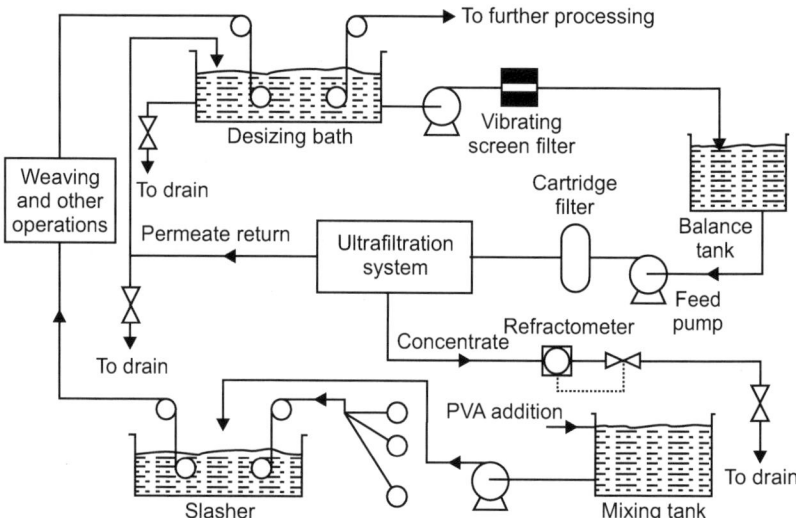

Figure 9.1: PVA recovery system.

9.3 Caustic recovery

Mercerising is a preparation step of cotton and cotton blends which uses a concentrated solution of sodium hydroxide (more than 20%). The recovery of caustic in this step is very practical since mercerising is a continuous operation which makes the characteristics of its waste stream are fairly constant. A good recovery system can recover up to 98% of the caustic. In another type of mercerising, the fabric is treated with liquid ammonia. The ammonia is captured as gas, recovered and reused. The benefits of caustic recovery are a reduced alkalinity of the wastewater and reduced chemical consumption.

The recovery of caustic from the mercerisation process is a common practice in the textile industry. Merceriser rinse water is normally recovered for evaporation when its concentration is above 2–3%. It is discharged to waste treatment when its concentration is below this level. Impurities from the fabric build up in the used caustic solution as caustic is removed. Eventually the solution wastestream must be discharged and the mercerisation process requires fresh caustic solution. An alternative to this procedure is to use an ultrafiltration membrane to filter the caustic water before the solution goes to the evaporator. The clarified and concentrated caustic solution is then ready for reuse and the consumption of caustic is significantly decreased. A diagram of a caustic recovery solution using ultrafiltration is shown in Fig. 9.2. The payback of invested capital was reported to be within 12 to 18 months. A return on investment such as this makes the process attractive to many plants.

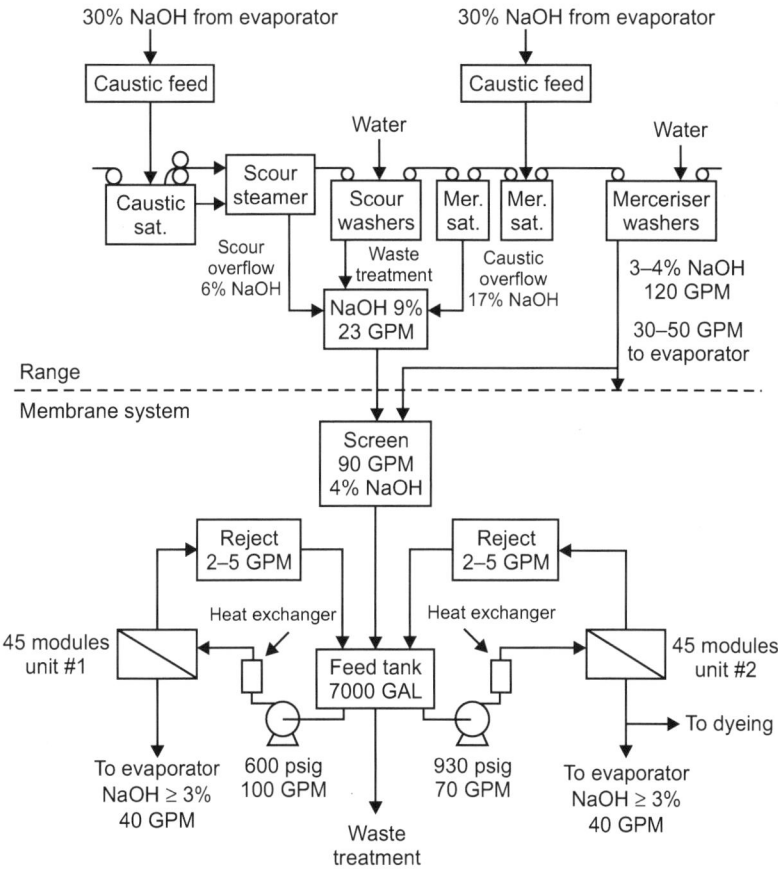

Figure 9.2: Caustic recovery system.

9.4   Indigo recovery

Recovery of indigo dye is an example of a system suited for the ultrafiltration process. The dye has a significant value and because of its deep blue colour would be readily visible in a receiving stream. If it is possible to recover the dye, pollution can be reduced and a savings in resources can be realised.

The recovery system is a multistage system with one feed pump and a bleed valve as shown in Fig. 9.3. Each stage automatically establishes a steady state concentration which becomes progressively higher as the concentration increases. The final stage reaches the maximum concentration of indigo for reuse. To minimise the cost of the membrane system, the indigo waste stream should be produced from countercurrent flow wash boxes adjusted to have a total flow nearing 25 gallons per minute. When the countercurrent flow is adjusted as such, it is possible to start recovery with an indigo concentration of near 800 ppm. This reduces the cost of the membrane system by half to a third below that normally found where the flow is 100 gallons per minute and the indigo concentration is 200 ppm.

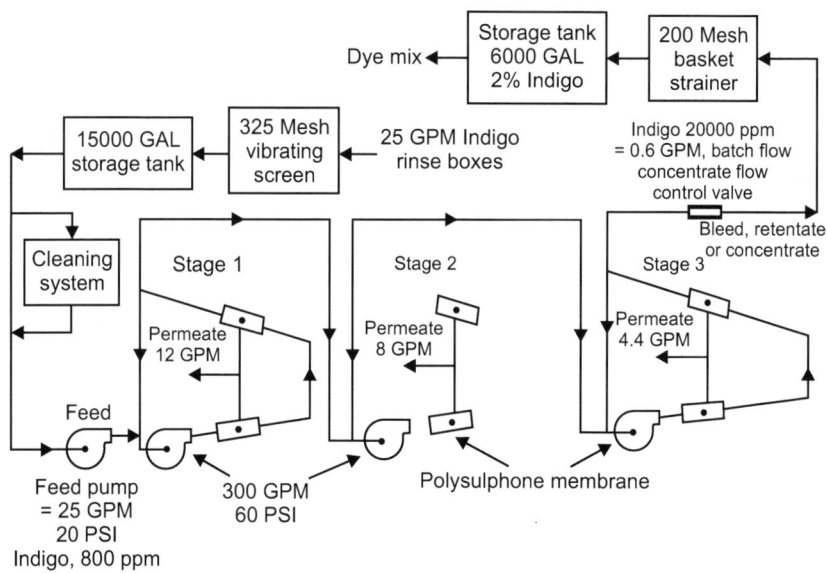

Figure 9.3: Indigo range recovery system.

The system is preceded by a 325-mesh vibrating screen used for lint removal from the waste stream before it enters the multistage membrane system. After the indigo is concentrated it is filtered through a 200-mesh basket strainer and stored in a holding tank capable of holding a four-day supply of concentrate. The ratio of concentration flow to feed flow generally ranges from 1:30 to 1:50. The clear filtrate is discharged to waste treatment containing the used chemicals from the dyeing process minus the indigo dye. The indigo recovery system is reported to have paid for itself in less than two years and operates with a minimum of problems.

9.4.1   Preparation

Fabric preparation is the most important step for gaining control of the dyeing and finishing operation. If the fabric is not consistently and uniformly prepared it will be difficult for the dyeing operation to make adjustments to correct for differences in fabric wet pickup. If the subsequent dyeing process is continuous, the problem of preparation nonuniformity will be severe because the fabric will wet out differently or unevenly. If the dyeing process is batchwise, dyeing conditions may have to be changed. All of these factors are well known and the textile industry is well aware of the need for good fabric preparation.

The normal procedure used to attain good preparation is to set the range conditions to those needed for difficult to prepare fabrics. These conditions are then used for temperature, water flow and chemical feeds and remain at these settings even though they may not be required but a fraction of the time. The process illustrated has no automatic process control and stream or water flow will vary as plant water and steam pressure vary. In this regard, fabric requirements may not be met when plant water and steam pressure are low. Changing to automatic control adjusts supply to need rather than a setting that may have nothing to do with the fabric being prepared at a specific time.

To recover chemicals and hot water from a process it is most important to control the process and know the exact needs for the process. The recovery process can then be designed to an optimum size and the holding capacity necessary for reuse can be designed properly.

Preparation range has three stages: Desizing, scouring and bleaching. When the fabric is to be mercerised it must be transferred to a separate range for mercerisation. The wash water from mercerisation is generally collected for evaporative recovery when it is near 3% NaOH or higher. When it is below 2% it is discharged to waste treatment. The peroxide washer wastewater may be used directly as feed to the caustic washer. When this is done it is possible to save $120,000 per year (As on year 2013). However, the need to pump the water and the retrofit costs have discouraged many textile plants from making the change. The total solids present in the peroxide washwater is generally less than 0.5% and should cause few problems for the caustic washer. One point of caution is the use of silicate stabilisers for the peroxide bleach. The stabilisers could interact with calcium or magnesium salts present in the natural cotton fibre and give a precipitate. In many cases, organic stabilisers are used which do not create a problem. Proper selection of stabilisers can enhance the success of peroxide washwater reuse.

The caustic washer contains less than 1% total solids. It can be used as a desize washer when the size is not recovered. This is a more difficult option but the potential to save an additional $120,000 (As on year 2013) makes the process attractive. The decrease in total wastewater flow not only saves on the cost of water, waste treatment and energy but can improve the biological treatment process used by most textile plants. When the plant recycles water, the water flow going to the waste treatment plant is reduced and the retention time available for biological treatment increased. This improves the biological waste treatment if no change is made in the volumetric capacity of the waste treatment system.

The overall potential for savings in caustic, water, steam and waste treatment could be over $300,000 per year (As on year 2013). Once demonstrated as practical, installation of a few pumps and one or two screen filters are the only requirements. When sensitive fabrics are processed, the range could automatically adjust flows to meet fabric requirements.

9.4.2   Beck, beam or jet dyeing

The wastewater from dyeing is more difficult to recycle than water from the previously discussed processes. This is because the colour of the dyeing wastewater continually changes. An exception to this is the indigo process where the colour is always the same. If sulphur dyes are not used of, if they are applied after the indigo dye, no contamination occurs in the indigo wastewater. With most piece dyeing operations, the target shades vary drastically. For this reason, it is difficult to separate the recovered dye in sufficient quantities to make the dye reuse practical. Dye recycling has been demonstrated as a successful recovery method. The major limitation to the practical use of dye recycling as a cost effective recovery method is that the quantity of dye recycled must be sufficient to make it worthwhile.

The water will not always be heated but most of the process steps of concern for potential energy savings operate faster and more effectively with hot water.

Thus, it is possible to recover over $100,000 per year (As on year 2013) from energy and water savings. In this case a membrane must be used to remove soluble dyes and must withstand the temperatures of the wastewater which can approach 200°F.

9.4.3   Continuous dyeing

Recovery of dyes and hot water from the continuous dyeing process was illustrated with the recovery of indigo. The indigo process operates continuously to dye warp yarns which are woven into denim fabric. When dyes and auxiliary chemicals are used to dye large quantities of fabrics, the dyes can be recovered and reused. When the quantity of dye used is small, the recovery of dye is not practical. However, in the case evaluated on a continuous dye range the hot water (>180°F) and waste treatment savings amounted to $200,000 to $300,000 annually (As on year 2013) depending on the price of oil and the waste treatment process. At this savings a recovery system could pay for itself in one to two years.

9.4.4   Potential for dyebath reuse

Dyebath reconstitution and reuse is an active process due to cost reduction, energy savings and pollution reduction. Dyebath reuse has been used for many dyes and materials. This section will discuss the procedure and will give examples where the technique has been used successfully.

Batch dyeing is inefficient in the 'use of chemicals, energy and water. The amount of auxiliary chemicals used varies from a few % to over 100% on the weight of the fabric. Most of these chemicals do not absorb into the fabric and increase the waste load of the mill's effluent. Dye quantities are often only a few % of the weight of the fabric. By reconstituting and reusing dyebaths, the efficiency of batch dyeing can be increased, and the use of chemicals, water and energy can be reduced significantly.

This could be feasible if the dyebath can be stored until the same material is dyed with the same dye formula or if it can be reused to dye the same material to a different shade.

Bergenthal and others suggested the following procedure for dyebath reconstitution:

Store the exhausted dyebath: The exhausted dyebath can be pumped into a holding tank where it is analysed and reconstituted. In the meantime the fabric is rinsed in the dye machine. The same can be achieved with two identical dye machines. One machine is preparing the yarn or fabric for dyeing while the other machine is dyeing the material. After dyeing, the dye solution of the second machine is pumped to the first machine for analysis and reconstitution. The second machine will be after-rinsing the fabric while machine 1 is in its dye cycle. Another alternative is to remove the fabric from the dye machine after dyeing and leave the exhausted dyebath in the dye machine for analysis and reconstitution. This eliminates the need for holding tanks.

Analyisis of the dyebath for residential chemicals: Dyestuff that is not exhausted from the dyebath can be measured by a spectrophotometer. If the dyebath is cloudy, extraction methods should be used. Most auxiliary chemicals will not be removed from the dyebath. The makeup quantity can be estimated or can be determined analytically. According to Smith, estimation of the losses is, in most cases, sufficient. Tincher and others have developed a computer programme that can help to determine the amount of auxiliary chemicals and dyes needed to reconstitute the dyebath.

Reconstituting the dyebath: In this step the necessary quantities of dyestuff, auxiliary and specialty chemicals are added to the exhausted dyebath. Water is also added to the bath to makeup for evaporation and the volume carried off by the fabric.

Reuse the dyebath: After the addition of the necessary chemicals and water, the temperature of the dyebath is raised to the desired temperature. Considerable time and energy are saved since the temperature of the reconstituted dyebath is higher than the temperature of the mill's source water.

Potential for dyebath reuse: The dyes applied to and the dyeing procedure used, depend largely on fibre characteristics. Also, auxiliary chemicals and specialty chemicals added to the dyebath vary with the fibre. It is therefore no surprise that the results of dyebath reuse vary. Cook showed that batch dyeing systems can be adapted for dyebath reuse. He came to this conclusion through several case studies. All case studies were successful and resulted in significant annual savings. The dyebath reconstitution can be practiced for many fabrics, dye classes and dye machines. Some case studies conducted included dyeings in which the shades achieved were different from the previous one.

Tincher and others developed a system to decolourise and reuse dyebaths with ozone and/or singlet oxygen. They concluded that acid and basic dyes can be destroyed with ozone treatment. The destruction of disperse dyes requires a much higher quantity of ozone. The auxiliary chemicals present in the dyebath are more resistant to ozone treatment. The lab dyeings showed that polyester carpet can be dyed with ozone decolourised dyebaths. In this case, it is necessary to treat the dyebath with a reductant before reuse. The carpets were of acceptable quality. Nylon carpets dyed in ozone treated dyebaths are also of acceptable quality.

Limitations

The success of dyebath reuse depends upon the type of dye and fabric. The easiest dyes to be reused for a limited number of dyecycles due to the buildup of impurities. Chemicals used in pretreatment steps and impurities from the fabric can accumulate in the dyebath. 'Impurities can also be present in auxiliary chemicals added to the dyebath. Some of these impurities can retard the dyeing process or can cause spotting. Cook and Tincher reported in 1978 that, when dyeing pantyhoses with disperse dyes, a dulling of the shade occurred in the tenth dyecycle. According to these researchers, the number of dyecycles can be increased by passing part of the exhausted dyebath through a ultrafiltration unit. This will lower the buildup of impurities.

After treatment of the fabric with chemicals is required when dyeing with reactive, vat and sulphur dyes. As a result, storage equipment is required to hold the exhausted dyebath when the dyed fabric is after treated. This increases the equipment cost and the quantity of water required for cleaning. Dyebath analysis is difficult when using reactive dyes because spectrophotometry can not differentiate between hydrolyzed and intact dyes.

9.4.5   Size recovery

Sizing chemicals are used in large amounts in mills processing woven fabric. In fact, they represent the largest group of chemicals used in the textile industry. The recovery of these chemicals has great pollution prevention opportunities. Some materials, like starch, are degraded which makes their recovery impossible. This is why some mills change to synthetic sizing agents like polyvinyl alcohol and carboxymethyl cellulose. Synthetic sizes pass the desizing process unchanged and can be recovered by ultrafiltration (UF) systems. The recovery of size is mostly only practiced in vertically integrated mills. Mills that buy woven fabric do not invest in size recovery equipment since they do not have the benefit of the recovered material. On the other side, synthetic sizes are more expensive than starch-like chemicals. Mills that weave the yarn but do not desize it after the weaving process, do not buy the more expensive synthetic sizes. This is one example of a situation where an arrangement between two mills resulted in benefits for both parties.

Polyvinyl alcohol can also be reclaimed by vacuum extraction. Currently, vacuum extraction is widely used to remove water from fabric before drying. Perkins reported that drying requirements can be lowered by more than 50% on some fabrics by extraction of unbound water before drying. He also investigated the recovery of PVA by vacuum extraction. This was done by either saturating the fabric with water in a desize saturator or by spraying the fabric. Afterwards, the cloth passed through a vacuum extractor.

The recoverability of the PVA depended on its viscosity and water solubility. The temperature of the water was also an important factor. He reported a recovery of 53% of the size from 50/50 polyester/cotton at a vacuum of 15 inches of mercury.

9.5   Efficient methods for the removal of chromium from textile effluents

Chromium is a metal that exists in several oxidation or valence states, ranging from Cr(-II) to Cr(+VI). Many effluents from leather industries, textile, dye industries, cement industries give away these toxic heavy metals that disturb our eco system. This section includes the methods to remove this hexavalent chromium. Chemical precipitation, adsorption and biosorption, reverse osmosis, ion exchange, electrodialysis and photocatalysis.

Chromium compounds are very stable in trivalent state and occure naturally in this state in ores such as ferrochromite or chromite ore. Chromium III is an essential nutrient for maintaining blood glucose levels. Hexavalent chromium is the second most stable compound that rarely occurs in nature. It is generally toxic and man made. It is used in many industrial application for its anti corrosive properties and it is used in electroplating.

The sources of chromium are aircraft painting, leather tanning, textile manufacturing, dyeing and cement industry. But it is hazardous to health. It is erosive to stomach, causes hemorrhage. Direct eye contact with chromic acid or chromate dusts can cause permanent eye damage. Hexavalent chromium can irritate eye, nose and throat. It can also cause ulcers, allergic reaction, skin rashes and is carcinogenic. It even causes death.

9.5.1   Methods of removal of chromium

Chemical precipitation

Formation of seperable solid substance from a solution, either by converting the substance into an insoluble form or by changing the composition of the solvent to diminish the solubility of the substance in it. The removal of chromium can be accomplished by the addition of ferrous sulphate and lime. Ferrous ion first reduces hexavalent chromium to trivalent chromium by simultaneous oxidation of ferrous ion to ferric. The resulting forms can be precipitated as hydroxides by lime. Chromium is precipitated as hydroxide:

$$Cr^{3+} + (OH)^- \rightarrow Cr(OH)^3$$

The process requires addition of other chemicals, which finally leads to the generation of a high water content sludge, the disposal of which is cost intensive. Its efficiency is affected by low pH and the presence of other salts (ions) and is ineffective in removal of the metal ions at low concentration.

Adsorption and biosorption

Adsorption is defined as accumulation of liquid or gas phase on the surface of a solid phase. The material that absorbs is called adsorbent and the substance getting adsorbed is called adsorbate. Adsorption may be physical adsorption or chemical adsorption or a combination of both (Fig. 9.4). Sometimes the adsorbent materials have negatively charged ligands that can form complexes with metal ion via electrostatic interactions. If we have to remove soluble material from the solution phase, but the material is neither volatile nor biodegradable, we often employ adsorption processes. Biosorption is a property of certain types of inactive, dead, microbial biomass to bind and concentrate heavy metals from even very dilute aqueous solutions. It is particularly the

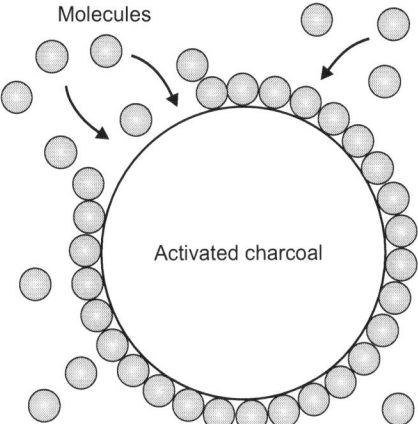

Figure 9.4: Adsorption of chromium by using activated charcoal.

cell wall structure of certain algae, fungai and bacteria which was found responsible for this phenomenon. The advantages of biosorption process are that these are cost effective, technically feasible and eco-friendly. This method suffers from low adsorption capacity and less intensity of biosorption.

Reverse osmosis

Reverse osmosis is filteration process that is often used for water. It works by using pressure to force a solution through a membrane, retaining the solute on one side and allowing the pure solvent to pass to the other side. This is the reverse of the normal osmosis process, which is the natural movement of solvent from an area of low solute concentration, through a membrane, to an area of high solute concentration when no external pressure is applied. A semi permeable membrane, like the membrane of a cell wall or a bladder, is selective about what it allows to pass through, and what it prevents from passing. These membranes in general pass water very easily because of its small molecular size; but also prevent many other contaminants from passing by trapping them. Water will typically be present on both sides of the membrane, with each side having a different concentration of dissolved minerals. Since the water with the less concentrated solution seeks to dilute the more concentrated solution, water will pass through the membrane from the lower concentration side to the greater concentration side. Eventually, osmotic pressure will counter the diffusion process exactly, and an equilibrium will form. The semi permeable membrane can be fabricated by a variety of materials in a way to support a high transmembrane pressure. Generally membranes made up of polyamide are used for the treatment of chromium containing effluent. The reverse osmosis technique has been successfully used in the treatment of electroplating rinse waters, not only to meet effluent discharge standards, but also to recover concentrated metal salt solutions for reuse. Its main demerits are high priced equipment and/or expensive monitoring system, high energy requirement, sludge generation.

Ion exchange

It is based on the exchange of cations or anions on synthetic resins with essential characteristics of its regeneration after the elution of ions. Resins are classified based on the type of functional group they contain:

Cation exchangers: Strongly acidic-functional groups derived from strong acids, e.g., $R-SO_3H$ (sulphonic). Weakly acidic functional groups derived from weak acids, e.g., $R-COOH$ (carboxylic).

Anionic exchangers: Strongly basic-functional groups derived from quaternary ammonia compounds, R-N-OH. Weakly basic- functional groups derived from primary and secondary amines, $R-NH_3OH$ or $R-R'-NH_2OH$.

About 100% removal of Cr (VI) was achieved in the studies. Its advantages over other processes are the recovery of the metal's value, high selectivity, less sludge volume produced and the ability to meet strict discharge specifications. Its limitations are high operating costs compared to other treatment systems. There can be incomplete removal of the chromium from the salt solution

Electrodialysis (ED)

Electrodialysis is an electro membrane process in which ions are transported through ion permeable membranes from one solution to another under the influence of a potential gradient. The electrical charges on the ions allow them to be driven through the membranes fabricated from ion exchange polymers. Applying a voltage between two end electrodes generates the potential field required for this. Since the membranes used in electro dialysis have the ability to selectively transport ions having positive or negative charges.

The ion permeable membranes used in eletrodialysis are essentially sheets of ion exchange resins. They usually contain other polymers to improve mechanical strength and flexibility. The resin components of a cation exchange membrane would have negatively charged groups chemically attached to the polymer chains. Polymer chain forms anion permeable membranes, which are selective to transport of negative ions, because the fixed $-NR^{3+}$ groups repel positive ions. The recovery percentage of chromium is quite good, the chromium concentration is not high enough to be cycled to the tanning process. Other problems with electrodialysis are high capital and operating costs involved and the requirement of highly trained human resources. The fouling and scaling of membranes is another drawback which can be controlled to an extent by employing flushing step.

Photo catalysis

Photo catalysis over a semiconductor oxide such as TiO_2 is initiated by the absorption of a photon with energy equal to, or greater than the band gap of the semiconductor producing electron hole pairs.

$$TiO_2 \rightarrow e^-_{cb} (TiO_2) + h^+_{vb} (TiO_2)$$

Oxidation of water by the hole produces the hydroxyl radical. Similarly O_2 radical are also formed. OH radicals rapidly attack pollutants in solution. The oxidation pathway is not yet fully understood. But OH radical can be formed in two different manners.

$$TiO_2(h_{vb}^+) + H_2O_{ads} \rightarrow TiO_2 + HO_{ads} + H^+$$

$$TiO_2(_{hvb}^+) + HO_{ads} \rightarrow TiO_2 + HO_{ads}$$

Cr(VI) will be reduced to Cr(III) and precipitated out. Photo catalysis has large capability for the removal of trace metals. The drawback of this method is that of being slow compared with traditional methods but it has the advantage not leaving toxic by product or sludge to be disposed.

Result and discussion

500 ml of effluent water is taken for each analysis, the methods are considered in the order for the removal of maximum chromium percentage. The first method, the chemical precipitation shows that it removes 69% of chromium and the second method, adsorption and biosorption to remove 72% of chromium. The further study shows from that reverse osmosis method the chromium removal of 76% is the highest chromium removal method among ion exchange, electrodialysis and photocatalysis method which is discussed in (Table 9.1).

Table 9.1: Removal of chromium using various method in %.

Method	*% present*	*% removed*
Chemical precipitation	80	69
Adsorption and biosorption	80	72
Reverse osmosis	80	76
Ion exchange	80	75
Electrodialysis	80	74
Photo catalysis	80	68

To sum up, each method has its own merits and limitation. The versatility, simplicity, cost effectiveness and technical feasibility are a few factors that must be considered while selecting a particular method.

High cost and technical complication are the problem associated with reverse osmosis and electrodialysis. Ion exchange is also comparatively costly whereas chemical precipitation leads to sludge generation and involve high capital costs. Photo catalysis process is still in the developmental stage. Adsorption and biosorption are found to be technically uncomplicated as well as economical but the desorption studies on the adsorbents need to be carried out before going for large scale applications.

Recycling and conservation of water

10.1 Introduction

The textile Industry is in no way different than other chemical industries, which causes pollution of one or the other type. The textile industry consumes large amount of water in its varied processing operations. In the mechanical processes of spinning and weaving, water consumed is very small as compared to textile wet processing operations, where water is used extensively. Almost all dyes, specialty chemicals and finishing chemicals are applied to textile substrates from water baths. In addition, most fabric preparation steps, including desizing, scouring, bleaching and mercerising use aqueous systems. According to USEPA a unit producing 20,000 lb/day of fabric consume 36,000 litres of water.

In textile wet processing, water is used mainly for two purposes. Firstly, as a solvent for processing chemicals and secondly, as a washing and rinsing medium. Apart from this, some water is consumed in ion exchange, boiler, cooling water, steam drying and cleaning.

Textile Industry is being forced to consider water conservation for many reasons. The primary reasons being the increased competition for clean water due to declining water tables, reduced sources of clean waters and increased demands from both industry and residential growth, all resulting in higher costs for this natural resource. Water and effluent costs may in the more common cases, account for as much as 5% of the production costs.

10.2 Water usage

Water usage at textile mills can generate millions of gallons of dye wastewater daily. The unnecessary usage of water adds substantially to the cost of finished textile products through increased charges for fresh water and for sewer discharge. The quantity of water required for textile processing is large and varies from mill to mill depending on fabric produce, process, equipment type and dyestuff. The longer the processing sequences, the higher will be the quantity of water required. Bulk of the water is utilised in washing at the end of each process. The processing of yarns also requires large volumes of water. The water usage of different purposes in a typical cotton textile mill and synthetic textile processing mill and the total water consumed during wet process is given in Table 10.1 and Table 10.2 respectively.

Table 10.1: Water usage in textile mills.

Purpose	% water use	
	Cotton textile	Synthetic textile
Steam generation	5.3	8.2
Cooling water	6.4	–
Dematerialised or RO water for specific purpose	7.8	30.6
Process water	72.3	28.3
Sanitary use	7.6	4.9
Miscellaneous and fire fighting	0.6	28.0

Table 10.2: Total water consumed during wet processing.

Process	% water consumed
Bleaching	38%
Dyeing	16%
Printing	8%
Boiler	14%
Other uses	24%

Wide variation is observed in consumption mainly due to the use of old and new technologies and difference in the processing steps followed types of machines used. Every textile processor should have knowledge of the quantity of water used for processing.

The volume of water required in litres/1000 kg of product for each process is given in Table 10.3.

Table 10.3: Water requirements for cotton textile wet finishing operations.

Process	Requirements in litres/1000 kg of product
Sizing	500–8200
Desizing	2500–21000
Scouring	20000–45000
Bleaching	2500–25000
Mercerising	17000–32000
Dyeing	10000–300000
Printing	8000–16000

Water requirements for synthetic textiles wet finishing operations are shown in Table 10.4.

Table 10.4: Water requirements for synthetic textiles wet finishing operations.

Process	*Requirements in litre/1000 kg of product*				
	Rayon	*Acetate*	*Nylon*	*Acrylic/ modacrylic*	*Polyester*
Scouring	17000-34000	25000-84000	50000-67000	50000-67000	25000-42000
Salt bath	4000-12000	–	–	–	–
Bleaching	–	33000-50000	–	–	–
Dyeing	17000-34000	34000-50000	17000-34000	17000-34000	17000-34000
Special finishing	4000-12000	24000-40000	32000-48000	40000-56000	8000-12000

Figure 10.1 show the scheme for washing tank in textile industry.

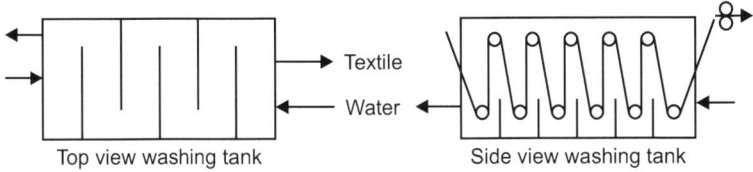

Figure **10.1:** Scheme of a washing tank in textile industry.

10.2.1   Washing efficiency

Washing process is characterised by its washing efficiency that is the amount of the compound that is removed divided by the total amount that could have been removed. Washing efficiency is not directly dependent on the amount of water used but is a function of:

1. Temperature.
2. Speed of fabric in the washing range.
3. The properties of the fabric.
4. The properties of the washing range.

Temperature

The temperature is important in washing because the temperature influences:

1. The viscosity of water: At a lower viscosity water can better penetrate through the fabric and washing will become more effective.
2. The affinity of compounds: At a higher temperature the affinity decreases and results in a better washing away of the unwanted components.
3. Migration of the components from the inner fibre to the water around the fibre: This migration is important for the total time the washing process will take.

Speed of fabric in the washing range

The speed of fabric in the washing range determines the amount of water that is hanged in the fabric by passing a roller in the washing compartment. That is the liquor that was in the fabric before passing the roller with a high concentration of the unwanted components that is replaced by the washing liquor with a low concentration of the unwanted components.

Properties of the fabric

The properties of fabric influence the washing effectiveness by the amount of water that can be pressed through the fabric during washing. The openness of the fabric as well as the openness of yarn determines the length of the way the unwanted component has to migrate to the fluid that can be exchanged in the washing process. In the washing process generally only very little water from the pores between the fibres is exchanged when the yarns are strongly twisted this will be practically zero. As migration is a very slow process it will take much longer for all the components to be washed out. The same holds more or less for thicker yarns and heavy weight fabrics.

Properties of the washing range

The effectiveness of the washing range is determined by the number of washing tanks, the number of compartments in each tank, the diameter of the roller and the way the fabric is led through the washing range. The washing effectiveness can be improved by placing rollers on top of the top-rollers. This squeeze off the excess water in the fabric and a better exchange of washing liquid will be realised. Also at high speed these top rollers will prevent water to be taken with the fabric to the next compartment.

Water conservation and reuse

Water is expensive to buy, treat and dispose. If the industry does not have water conservation programme, its pouring money downs the drain. Now, water conservation and reuse are rapidly becoming a necessity for textile industry. Water conservation and reuse can have tremendous benefits through decreased costs of purchased water and reduces costs for treatment of wastewaters. Prevention of discharge violations as a result of overload systems can be a significant inducement for water conservation and reuse. By implementing water conservation and reuse programmes, the decision to expand the treatment facilities can be placed on hold and the available funds can then be used for expansion or improvements to process equipment.

The first step in developing a water conservation and reuse programme is to conduct a site survey to determine where and how water being used. It would be extremely helpful to develop a spreadsheet and/or diagram of the water usage with specific details as shown below:

1. Location and quantity of water usage.
2. Temperature requirements.
3. Water quality requirements, i.e., pH, hardness and limitations on solid content, must meet clean water standards, etc.
4. Any special process requirements.

10.2.2   Water conservation measures

Water conservation measures lead to:

1. Reduction in processing cost.
2. Reduction in wastewater treatment cost.
3. Reduction in thermal energy consumption.
4. Reduction in electrical energy consumption.
5. Reduction in pollutants load.

Water conservation significantly reduces effluent volume. A water conservation programme can cut water consumption by up to 30% or more and the cost savings can pay for the required materials in a very short time. Since the average plant has a large number of washers, the savings can add up to thousands of rupees per year.

Other reasons for large effluent volume is the choice of inefficient washing equipment, excessively long washing circles and use of fresh water at all points of water use.

The equipment used in a water conservation programme is relatively inexpensive, consisting in most cases of valves, piping, small pumps and tanks only. The operating costs for these systems are generally very low. Routine maintenance and, in some cases, electricity for the pumps, would be the major cost components.

The payback period for a water conservation system will vary with the quantity of water saved, sewer fees and costs for raw water and wastewater treatment. In addition to the direct cost savings, a water conservation programme can reduce the capital costs of any required end-of-pipe wastewater treatment system. Personnel from textile industry need to be aware of water conservation potential so they can help their organisation realise the benefits.

10.2.3   Water conservation methods for textile mills

Numerous methods have been developed to conserve water at textile mills. Some of the techniques applicable to a wide variety of mills are discussed.

Good housekeeping

A reduction in water use of 10 to 30% can be accomplished by taking strict housekeeping measures.

A walk through audit can uncover water waste in the form of:

1. Hoses left running.
2. Broken or missing valves.
3. Excessive water use in washing operations.
4. Leaks from pipes, valves and pumps.
5. Cooling water or wash boxes left running when machinery is shut down.
6. Defective toilets and water coolers.

Good housekeeping measures often carried out without significant investments, but leading to substantial cost savings and the saving of water, chemicals and energy.

Good housekeeping measures are essential for a company, which is critical about its own behaviours. Implementing the following can make significant reductions in water use:

1. Minimising leaks and spills.
2. Plugging leakages and checks on running taps.
3. Installation of water meters or level controllers on major water carrying lines.
4. Turn off water when machines are not operating.
5. Identifying unnecessary washing of both fabric and equipment.
6. Training employees on the importance of water conservation.

10.2.4   Water reuse

Water reuse measures reduce hydraulic loadings to treatment systems by using the same water in more than one process. Water reuse resulting from advanced wastewater treatment (recycle) is not considered an in-plant control, because it does not reduce hydraulic or pollutant loadings on the treatment plant.

Reuse of certain process water elsewhere in mill operations and reuse of uncontaminated cooling water in operations requiring hot water result in significant wastewater discharge reductions.

Examples of process water reuse include the following:

Reuse of water jet weaving wastewater

The jet weaving wastewater can be reused within the jet looms. Alternatively, it can be reused in the desizing or scouring process, provided that in-line filters remove fabric impurities and oils.

Reuse of bleach bath

Cotton and cotton blend preparation are performed using continuous or batch processes and usually are the largest water consumers in a mill. Continuous processes are much easier to adapt to wastewater recycling/reuse because the waste stream is continuous, shows fairly constant characteristics and usually is easy to segregate from other waste streams.

Waste stream reuse in a typical bleach unit for polyester/cotton blend and 100% cotton fabrics would include recycling j-box and kier drain wastewater to saturators, recycling continuous scour wash water to batch scouring, recycling washer water to equipment and facility cleaning, reusing scour rinses for desizing, reusing mercerises wash water or bleach wash water for scouring.

Preparation chemicals, however, must be selected in such a way that reuse does not create quality problems such as spotting.

Batch scouring and bleaching are less easy to adapt to recycling of waste streams because streams occur intermittently and are not easily segregated. With appropriate holding tanks, however, bleach bath reuse can be practiced in a similar manner to dye bath reuse and several pieces of equipment are now available that has necessary holding tanks.

Reuse of final rinse water from dyeing for dye bath make-up

The rinse water from the final rinse in a batch dyeing operation is fairly clean and can be used directly for further rinsing or to make up subsequent dye baths. Several woven fabric and carpet mills use this rinse water for dye bath make-up.

Reuse of soaper wastewater

The coloured wastewater from the soaping operation can be reused at the back grey washer, which does not require water of a very high quality. Alternatively, the wastewater can be used for cleaning floors and equipment in the print and colour shop.

Reuse of dye liquors

The feasibility of dye liquor reuse depends on the dye used and the shade required on the fabric or yarn as well as the type of process involved. It has already been applied whilst disperse dyeing polyester, reactive dyeing cotton, acid dyeing nylon and basic dyeing acrylic, on a wide variety of machines. However, commission dyeing where the shades required are much more varied and unpredictable would make the reuse of dye liquor difficult. But, given the right conditions dye liquor could be reused up to 10 times before the level of impurities limits further use.

Reuse of cooling water

Cooling water that does not come in contact with fabric or process chemicals can be collected and reused directly. Examples include condenser-cooling water, water from water-cooled bearings, heat-exchanger water and water recovered from cooling rolls, yarn dryers, pressure dyeing machines and air compressors. This water can be pumped to hot water storage tanks for reuse in operations such as dyeing, bleaching, rinsing and cleaning where heated water is required or used as feeding water for a boiler.

Reusing wash water

The most popular and successful strategy applied for reusing wash water is counter-current washing. The counter-current washing method (Fig. 10.2) is relatively straight forward and inexpensive. For both water and energy savings, counter-current washing is employed frequently on continuous preparation and dye ranges.

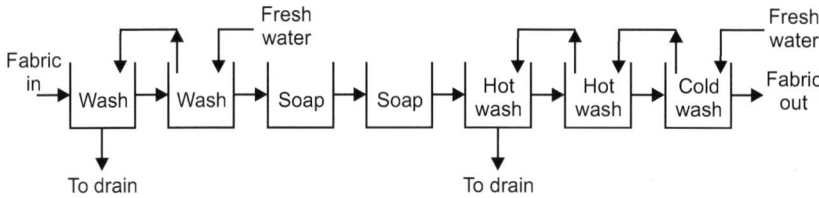

Figure 10.2: Recommended counter-current flow of washing on a soaper range.

Clean water enters at the final wash box and flows counter to the movement of the fabric through the wash boxes. With this method the least contaminated water from the final wash is reused for the next-to-last wash and so on until the water reaches the first wash stage, where it is finally discharged. Direct counter-current washing is now generally built into the process flow sheet of new textile mills. It is also easy to implement in existing mills where there is a synchronous processing operation.

Use of automatic shut-off valves

An automatic shut-off valve set to time, level, or temperature controls the flow of water into a process unit. One plant estimated that a reduction in water use of up to 20% could be achieved with thermally controlled shut-off valves.

Use of flow control valves

A flow or pressure-reduction valve can significantly reduce the quantity of water used in a wash or clean-up step. These valves are particularly useful in cleaning areas where operators are not always aware of the need for water conservation.

10.2.5 Flocculation of clean water of pigment printing

A rotary screen printer uses as much water as a continuous washing range. All this water is used to wash the belt, to rinse the pipes and pumps and to clean the screens and squeegees. The water does not come in contact with the fabric. When only pigments are used for printing, it is relatively easy to coagulate the pigments and let them settle. The result is the clean water, which can be used for cleaning purposes.

Use single stage of processing

Knitted fabric process combined bleaching/scouring and dyeing giving considerable saving in water. The scouring and bleaching process takes place for 10–20 minutes and without draining the bath the dyeing is carried out without any loss of depth of colour value. In some cases the finishing process can also be carried out along with the dyeing process.

Use of low material to liquor ratio systems

Different types of dyeing machinery use different amounts of water. Low liquor ratio dyeing machines conserve water as well as chemicals and also achieve higher fixation efficiency but the washing efficiency of some types of low liquor ratio dyeing machines, such as jigs, is inherently poor; therefore, a correlation between liquor ratio and total water use is not always exact. Typical liquor ratios for various types of dyeing machines are given in Table 10.5.

Table 10.5: Liquor ratio in different dyeing machines.

Dyeing machine	Typical liquor ratio/goods at time of dye application
Continuous	1:1
Winch	15:1–40:1
Jet	7:1–15:1
Jig	5:1
Beam	10:1
Package	10:1
Beck	17:1
Stock	12:1
Skein	17:1

10.2.6 Water conservation measures in dyeing equipment

Washing and rinsing are both important for reducing impurity levels in the fabric to pre-determined levels. Water and wastewater treatment prices are increasing, the optimisation of water use pays dividends. One possible option is to reduce rinse water use for lighter shades. Here are some successful water reduction projects in batch and continuous operations.

Winch dyeing: Dropping the dye bath and avoiding overflow rinsing could reduce water consumption reduced by 25%.

High and low: Replacing the overflow with Pressure jet dyeing batch wise rinsing can cut water consumption by approximately 50%.

Beam dyeing: About 60% of water preventing overflows during soaking and rinsing may reduce consumption. Automatic controls proved to be quite economical with a payback period of about four months.

Jig dyeing: A wide range of reductions ranging from 15 to 79% is possible by switching from the practice of overflow to stepwise rinsing. Rinsing using a spray technique is also effective.

Cheese dyeing: A reduction of around 70% is possible following intermittent rinsing.

Continuous operation: A 20–30% saving was realised by introducing automatic water stops. Counter-current washing proved to be the most effective method. Horizontal washing equipment delivered the same performance as two vertical washing machines, using the same amount of water.

10.3 Modern technologies can save water in textile industry

The textile industry is the backbone of many developing economies. It is also heavily reliant on water. New technologies and simple fixes are helping mills remain competitive while reducing their dependence on water and contributing to a better environment.

From farm to factory to your favourite store, your new cotton T-shirt required approximately 2650 litres of water to grow, produce and transport.

A substantial proportion of this water usage - 20% and more when using conventional methods - is used in just the dyeing phase of the process. Up to 100 litres of fresh water and very high amount of energy, is required to dye just one kilogram of cotton fabric. Much of this water ends up contaminated by the salt used to promote the absorption of the dye.

This salinated wastewater is difficult to treat and it cannot be safely consumed or used for irrigation and it is harmful to aquatic life.

These challenges are exacerbated in regions facing acute water scarcity. Unfortunately, many of the world's largest textile-producing nations-China, India, Bangladesh and Brazil, for example-are also those most vulnerable to water shortages.

10.3.1   On the cusp of change

The public in these textile-producing countries is becoming increasingly vocal about deteriorating water quality and the lack of sufficient clean water for homes and agriculture. People in the developed world are also beginning to demand that the garments and textile products they buy are eco-friendly.

This attitude change is putting pressure on brands and retailers to show that their supply chains are clean and transparent. Governments too have reacted by mandating more stringent environmental legislation and by more strictly enforcing their pollution laws.

Even so, the United Nations is warning that half the global population could be facing water shortages by 2030.

Taken together, all of these factors make reducing our use of water one of the most pressing challenges facing the textile industry.

10.3.2   Challenges ahead

A great deal can be done today to reduce the industry's reliance on water. Process and efficiency improvements can have some impact, but dramatic gains are needed.

10.3.3   Waterless dyeing

Waterless dyeing technology has been under development for several years. Recently, innovative sustainable products were developed by using recyclable carbon dioxide (supercritical CO_2) as the application medium to infuse colour into fabric instead of water. This completely eliminated the use of water in the textile dyeing process and would benefit the industry in years to come.

10.3.4   Digital textile printing

Digital textile printing has also come of age. It requires mills to invest heavily in digital printing machines and to retrain staff and use high-quality specialist inks, but it is now cost-effective for higher value fabrics. Digital printing allows mills to print an almost unlimited array of colours and complex patterns in short runs. It is also a very clean process that minimises waste and substantially reduces water and energy consumption.

Perhaps even more promising, however, are developments that help mills make dramatic savings without requiring substantial investment in new plant or equipment. These new innovation are a range of reactive dyes for cotton and cellulosic fibres using technology that assists in them getting absorbed by textile fibres more rapidly, using less salt during dyeing and less water during the wash-off process. These unique set of properties ensure high-quality results at much lower costs along with improved environment acceptability.

Even more dramatic gains are delivered against conventional dye house technologies, which many mills in developing nations still use. Statistically, this new and exciting technology could potentially save more than 820 billion litres of water per year, which equates to 1.3 litres of fresh water per person per day in the major Asian textile processing countries.

10.3.5 Accelerating the change

Textile industry is at an inflection point today, with environmental and competitive pressures demanding new approaches. Where can we go next?

To reduce our reliance on cotton, the industry will invest in less-thirsty alternative fibres, such as bamboo and other man-made fibres and recycled polyester. We can improve laundry equipment and detergents for industrial and home use. It is only necessary that we must lead the way with new dye technologies and cleaner processes that save water and energy.

However, substantially changing the supply chain in an industry as complex and global as ours will take time. Mills that serves high-volume, low-cost retailers have very small margins and we are already seeing some closing because they cannot meet tougher environmental regulations.

Changing the mindset of producers about water conservation will need to be an industry priority if we are to accelerate the pace of change. Simple changes like fixing leaks, installing sensors and water meters, collecting the monsoon rains and switching to new dyes can all pay big dividends. As an incentive for change, cost savings are hard to beat.

Reuse of process water: Large quantities of hot water are used to manufacture textiles, up to 200 T of water per ton of textiles. And this water is often simply flushed away, wasting both the heat and the water resource itself. Heat exchangers can be installed to reuse the energy from the hot rinse wastewater; these devices use the temperature of the out-going water to pre-heat the incoming water for the next hot rinse. In addition, hot water from condensed steam, valuable both for its high heat content and purity, can be reused in the process, again with simple measures that pay themselves back in less than six months. Finally, clean rinse water from the final rinse can be beneficially recycled and reused at the start of the rinse process. Taken together with improved maintenance and leak fixing, these opportunities can reduce water consumption by almost 25%.

Cogeneration

11.1 Introduction

Cogeneration is an energy production process involving simultaneous generation of thermal (e.g., process steam) and electric energy by using single primary heating source. Fuel saving is the major incentive for the use of cogeneration since all heat engine based electric power systems reject heat to environment. That rejected heat can frequently be used to meet all or part of the on-site or local thermal energy meets. Use of rejected heat usually has no effect on the amount of primary fuel used, yet it leads to saving in all or part of the fuel that would otherwise be used.

Generally, the primary energy form is thermal (steam) and the secondary form is either electrical or mechanical. The electrical or mechanical energy can be used internally to run company equipment and thus reduce the demand for utility power and the surplus electricity, if available, can be sold to the utilities. Such a system can reduce energy input to 10–30% of what is required by separate systems to produce the same outputs. Total system efficiency can approach 90%, a significant improvement over the 50–90% efficiency of many industrial boilers and 30–35% efficiency of electrical conversion when separate production is used.

As a result, this simultaneous efficient production of two energy forms can significantly reduce total operating costs in many instances, even after paying for the increased capital costs.

11.2 Prime movers for cogeneration

11.2.1 Steam turbine

Steam turbines are the most commonly employed prime movers for cogeneration applications. In the steam turbine, the incoming high pressure steam is expanded to a lower pressure level, converting the thermal energy of high pressure steam to kinetic energy through nozzles and then to mechanical power through rotating blades.

11.2.2 Gas turbine

The fuel is burnt in a pressurised combustion chamber using combustion air supplied by a compressor that is integral with the gas turbine. In conventional gas turbine gases enter the turbine at a temperature range of 900 to 1000°C and leave at 400 to 500°C. The very hot pressurised gases are used to turn a series of turbine blades and the shaft on which they are mounted, to produce mechanical energy. Residual energy in the form of a high flow of hot exhaust gases can be used to meet, wholly or partly, the thermal (steam) demand of the site. Waste gases are exhausted from the turbine at 450°C to 550°C, making the gas turbine particularly suitable for high-grade heat supply.

The available mechanical energy can be applied in the following ways:

1. To produce electricity with a generator (most applications).
2. To drive pumps, compressors, blowers, etc.

A gas turbine operates under exacting conditions of high speed and high temperature. The hot gases supplied to it must therefore be clean (i.e., free of particulates which would erode the blades) and must contain not more than minimal amounts of contaminants, which would cause corrosion under operating conditions. High-premium fuels are therefore most often used, particularly natural gas. Distillate oils such as gas oil are also suitable and sets capable of using both are often installed to take advantage of cheaper interruptible gas tariffs. LPGs and Naphtha are also suitable, LPG being a possible fuel in either gaseous or liquid form.

11.2.3 Reciprocating engine systems

This system provides process heat or steam from engine exhaust. The engine jacket cooling water heat exchanger and lube oil cooler may also be used to provide hot water or hot air. There are, however, limited applications for this.

As these engines can use only fuels like HSD, distillate, residual oils, natural gas, LPG, etc., and as they are not economically better than steam/gas turbine, their use is not widespread for co-generation. One more reason for this is the engine maintenance requirement.

11.3   Methods for cogeneration

Industrial cogeneration or electricity production is feasible by using following four techniques: (i) gas turbine topping cycle, (ii) steam turbine topping cycle, (iii) combined cycle and (iv) bottoming cycle.

11.3.1   Gas turbine topping cycle

In this method air is mixed with fuel in combustion chamber. The hot gases from the chamber are fed to a gas turbine, which produces electricity. The hot exhaust gases from the gas turbine are fed to wasted heat recovery boiler which generate power-steam at the desired temperature and pressure (Fig. 11.1).

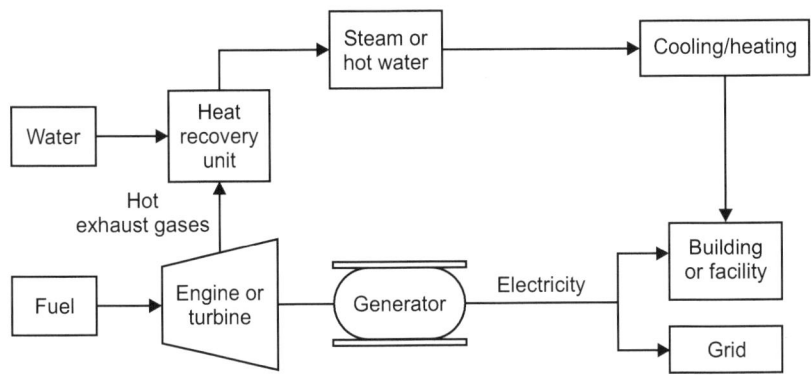

Figure 11.1: Gas turbine or engine with heat recovery unit.

11.3.2   Steam turbine topping cycle

In this method steam is generated at a pressure greater than from an industrial steam plant that produces power as a by-product or required for the process and then taken down to the desired pressure through back pressure turbine coupled to a generator. The steam for electric power generation can vary from 10% of total steam used in an industrial steam plant that produces power as a by-product to 70% or more (Fig. 11.2).

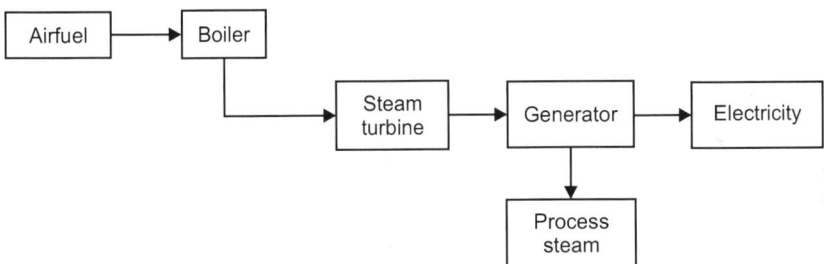

Figure 11.2: Steam turbine topping cycle.

11.3.3   Combined cycle

This plant consist of gas turbine generator, waste heat boiler and single steam turbine (Fig. 11.3). The compressor accepts inlet air from the atmosphere which is compressed to 120–160 psig. This burns along with fuel in proper air mixture to provide hot gases for the gas turbine which generates electricity. The heat recovery boiler converts gas turbine exhaust gases into steam energy. The boiler can be in fired, supplementary fired or mixed fired type. This provides steam to the back pressure turbine for generating electricity and for obtaining low-pressure process steam. This type of plant may provide benefits in situations

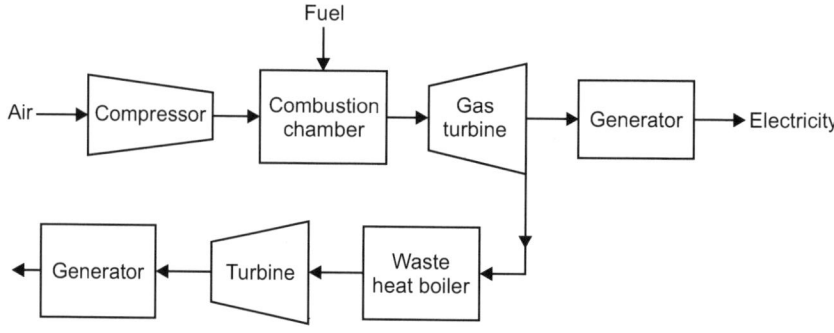

Figure 11.3: Combined cycle.

where the process steam requirement is variable and loss of power generation can be made up with conventional power plant. Combined cycle plants are capable of power generation at approximately 8000 Btu/kWh or 43% efficiency.

11.3.4   Bottoming cycle

In industrial process a significant amount of energy is released into the atmosphere at sufficiently high temperature. With the use of established technology it can further be utilised to produce useful work. The bottoming cycle working fluid can be either water or an organic fluid, which is used for lower temperature of waste gas. An attractive feature of bottoming cycle is the generation of electricity without fuel consumption (Fig. 11.4).

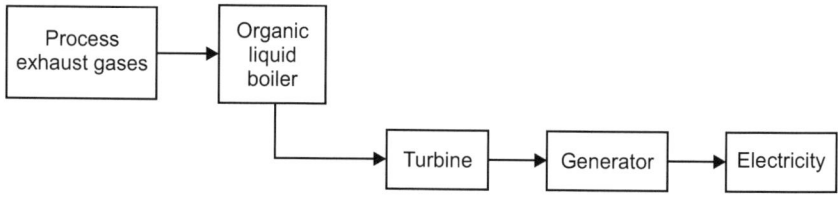

Figure 11.4: Bottoming cycle.

11.4   Energy conservation potential

The potential for conserving fuel through cogeneration depends on the user's demand for electricity and heat and also on the type of cogeneration system to be used. Steam turbine, combustion turbine and diesel engine systems will each produce a different ratio of electricity to heat output. Fuel conservation will depend on the ability of a cogeneration system to match the user's overall needs for each type of energy output and to cope with fluctuations in the individual demands. Steam is the primary form in which the heat output from a cogeneration system can be utilised. Steam is widely used in industry and to a somewhat lesser extent in buildings, to transport heat to a point of use. Almost half of all the industrial fuel burned is consumed in generating versatile, easy-to-use steam.

A large, well-designed, industrial boiler will convert 85–90% of the combustion heat of the fuel into steam. If this steam is used for heating and if after such use, the hot condensate water is returned to the boiler, the heat content of the steam can be very effectively used. However, when steam is employed to generate electricity in a modern, high-pressure, high-temperature, single-purpose plant, only 37–40% of the fuel combustion heat is converted to electricity. Most of the remaining heat is carried off by cooling when the exhaust steam passes through condensers at the end of the electricity-generating process. While rejecting this large (45–48%) proportion of a fuel's heat energy seems inefficient, the electricity produced is of high quality, with the capacity to do more work or generate higher temperature than could be obtained from the rejected low pressure steam.

While cogeneration systems improve the overall efficiency of fuel use, some of the gain is at the expense of electrical power generation. When steam is discharged from a turbine at pressures high enough for heating purposes, electrical power output is reduced proportionately. In a cogeneration system, producing by-product steam at a useful pressure of 50 psi, only about 15% of the input energy will come out as electricity and almost 70% will remain in the steam. Cogeneration systems based on diesel engines or combustion turbines can produce higher electricity to steam ratios. However, the ratio for a given system is relatively constant and must match the needs of the installation or it will restrict the energy conservation potential. If the actual demand ratio varies by more than a certain percentage from the design (expected) ratio for the system, it will usually be necessary to simply discard and waste the excess energy output.

Cogeneration systems may encounter highly variable demands for electricity and steam when serving end-uses such as mechanical work, electrical lighting, or heating. An industrial plant may have a peak steam or electricity demand during a particular shift every day. Consequently, demand for electricity may not coincide with a corresponding demand for the cogenerated heat. A useful fuel conserving cogeneration system must be able to produce electricity and heat (for example, steam) in proportions to match the load or array of uses served whenever required. In contrast, a prime attraction of conventional single-purpose systems for process steam, space heating, or electricity generation is that the fuel input and the operating characteristics of the system are controlled by only one output demand.

Different modes of cogeneration will vary in their output ratio of electricity and high- and low-temperature heat. A basic steam turbine cogeneration system produces only about 15% of electricity with 70% of heat energy in the exhaust steam. A combined cycle cogeneration system using a combustion turbine and an exhaust-heat boiler produces 30% of electricity with 47% of input heat in the exhaust steam. A diesel cogeneration system using an exhaust-heat boiler produces 36% of electricity with 27% of heat in the exhaust steam.

These ratios of electricity-to-steam are based on a steam pressure of three bars, which is somewhat low for average industrial use. For the higher pressures required in many steam systems, the electricity-to-steam ratio will be lower in the case of steam turbine cogeneration; but higher for the combustion turbine or diesel engine modes. However, in any system designed for higher electricity /steam ratios, the fuel consumption per unit of electricity also increases and the energy conservation potential is diminished. The overall efficiency of fuel use in a cogeneration system will, therefore, depend upon the type of system installed as well as the user requirements for electricity and high- or low-pressure steam from the system.

Cogeneration is currently favoured where an industrial plant can satisfy its needs for electricity and steam with a simple basic boiler and steam turbine system. Such plants will use 65–80% of their energy demand as low-pressure (~10 bars) steam and the remaining 20–35% as electricity. For example, these conditions are found in many paper plants; and the pulp and paper industry is a leader in industrial cogeneration. Approximately 30% of the electricity used in the pulp and paper industry is cogenerated with the steam production in paper mills. Another area where cogeneration has been actively employed and also integrated into the system, is the fertiliser industry. Almost all the energy requirements of the ammonia and urea plants are met by the steam generated through high pressure waste-heat boilers.

In general, industrial cogeneration systems are designed primarily to satisfy the heating requirements of an installation and electricity is considered only as a by-product. This electricity might be sold to the local utility grid or used internally. In either case, it replaces conventionally generated electricity in single purpose power plants. Cogenerated electricity, depending on the system used, is produced with only 40–70% of the fuel normally required in an efficient, single purpose power plant.

The potential for fuel conservation thus depends upon how much electricity as by-product can be cogenerated and this will depend upon both steam-or hot water heating demand of the user and the mode of cogeneration employed. For a given heating-energy output requirement, steam turbine cogeneration will yield electricity as a by-product with a maximum fuel saving equivalent to 30% of the energy in the steam output. In combustion turbine and diesel engine systems, the overall fuel savings can be equivalent to 70 and 85% of the respective steam energy output. However, because these two latter modes produce such high electricity-to-steam ratios, the fuel-saving per unit of electricity generated is actually less than that of the steam turbine mode.

The overall potential for fuel conservation using cogeneration depends upon the relationship between the size and characteristics of user demands for low temperature heat and for electricity and the mode of cogeneration employed. It is unlikely that cogeneration systems, which fulfil the operating and economic requirements of an individual user, will also produce maximum oil and gas savings from a national viewpoint. A user with a high electricity/low steam demand will not be served well by a cogeneration system with a low electricity /steam production ratio, even if that system provides fuel-cheap electricity.

Electricity production can ride industrial process steam production while attempting to make dual use of fuel. However, the electricity produced, or the steam cogenerated may not be available in the amount required or at the time it is needed. This conflict is but one of the factors hampering realisation of the full conservation benefits from cogeneration.

11.4.1   Topping maximum cogeneration potential

One of the prime requirements for tapping the optimum cogeneration potential in any plant is to have an integrated approach and energy conservation technology. An integrated approach involves a study of plant in it's totality and covers all energy system—from boiler, steam system with turbines, to electrical distribution, etc. In short, it involves a study and understanding of all sections of plant rather than the piecemeal approach of looking at one area here and another area there.

11.5   Factors influencing cogeneration choice

The selection and operating scheme of a cogeneration system is very much site-specific and depends on several factors, as described below.

11.5.1   Base electrical load matching

In this configuration, the cogeneration plant is sized to meet the minimum electricity demand of the site based on the historical demand curve. The rest of the needed power is purchased from the utility grid. The thermal energy requirement of the site could be met by the cogeneration system alone or by additional boilers. If the thermal energy generated with the base electrical load exceeds the plant's demand and if the situation permits, excess thermal energy can be exported to neighbouring customers.

11.5.2   Base thermal load matching

Here, the cogeneration system is sized to supply the minimum thermal energy requirement of the site. Stand-by boilers or burners are operated during periods when the demand for heat is higher. The prime mover installed operates at full load at all times. If the electricity demand of the site exceeds that which can be provided by the prime mover, then the remaining amount can be purchased from the grid. Likewise, if local laws permit, the excess electricity can be sold to the power utility.

11.5.3   Electrical load matching

In this operating scheme, the facility is totally independent of the power utility grid. All the power requirements of the site, including the reserves needed during scheduled and unscheduled maintenance, are to be taken into account while sizing the system.

This is also referred to as a 'stand-alone' system. If the thermal energy demand of the site is higher than that generated by the cogeneration system, auxiliary boilers are used. On the other hand, when the thermal energy demand is low, some thermal energy is wasted. If there is a possibility, excess thermal energy can be exported to neighbouring facilities.

11.5.4   Thermal load matching

The cogeneration system is designed to meet the thermal energy requirement of the site at any time. The prime movers are operated following the thermal demand. During the period when the electricity demand exceeds the generation capacity, the deficit can be compensated by power purchased from the grid. Similarly, if the local legislation permits, electricity produced in excess at any time may be sold to the utility.

11.6   Important technical parameters for cogeneration

While selecting cogeneration systems, one should consider some important technical parameters that assist in defining the type and operating scheme of different alternative cogeneration systems to be selected.

11.6.1   Heat-to-power ratio

Heat-to-power ratio is one of the most important technical parameters influencing the selection of the type of cogeneration system. The heat-to-power ratio of a facility should match with the characteristics of the cogeneration system to be installed.

It is defined as the ratio of thermal energy to electricity required by the energy consuming facility. Though it can be expressed in different units such as Btu/kWh, kCal/kWh, lb./hr/kW, etc., here it is presented on the basis of the same energy unit (kW).

Cogeneration is likely to be most attractive under the following circumstances:

1. The demand for both steam and power is balanced, i.e., consistent with the range of steam. Power output ratios that can be obtained from a suitable cogeneration plant.

2. A single plant or group of plants has sufficient demand for steam and power to permit economies of scale to be achieved.

3. Peaks and troughs in demand can be managed or, in the case of electricity, adequate backup supplies can be obtained from the utility company.

11.6.2   Quality of thermal energy needed

The quality of thermal energy required (temperature and pressure) also determines the type of cogeneration system. For a sugar mill needing thermal energy at about 120°C, a topping cycle cogeneration system can meet the heat demand.

11.6.3   Load patterns

The heat and power demand patterns of the user affect the selection (type and size) of the cogeneration system.

11.6.4   Fuels available

Depending on the availability of fuels, some potential cogeneration systems may have to be rejected. The availability of cheap fuels or waste products that can be used as fuels at a site is one of the major factors in the technical consideration because it determines the competitiveness of the cogeneration system.

11.6.5   System reliability

Some energy consuming facilities require very reliable power and/or heat; for instance, a pulp and paper industry cannot operate with a prolonged unavailability of process steam. In such instances, the cogeneration system to be installed must be modular, i.e., it should consist of more than one unit so that shut down of a specific unit cannot seriously affect the energy supply.

11.6.6   Grid dependent system versus independent system

A grid-dependent system has access to the grid to buy or sell electricity. The grid-independent system is also known as a 'stand-alone' system that meets all the energy demands of the site. It is obvious that for the same energy consuming facility, the technical configuration of the cogeneration system designed as a grid dependent system would be different from that of a stand-alone system.

11.6.7 Retrofit versus new installation

If the cogeneration system is installed as a retrofit, the system must be designed so that the existing energy conversion systems, such as boilers, can still be used. In such a circumstance, the options for cogeneration system would depend on whether the system is a retrofit or a new installation.

11.6.8 Electricity buy-back

The technical consideration of cogeneration system must take into account whether the local regulations permit electric utilities to buy electricity from the cogenerators or not. The size and type of cogeneration system could be significantly different if one were to allow the export of electricity to the grid.

11.6.9 Local environmental regulation

The local environmental regulations can limit the choice of fuels to be used for the proposed cogeneration systems. If the local environmental regulations are stringent, some available fuels cannot be considered because of the high treatment cost of the polluted exhaust gas and in some cases, the fuel itself.

11.7 Constraints on promoting cogeneration

While the extent of conservation potential through total energy systems has been well appreciated, as also the numerous advantages one would derive from the concept, there are several barriers to cogeneration, which restrict the adoption of such systems by the industries. The identified cogeneration market barriers are discussed briefly:

1. Regulatory uncertainty: The key regulations, which qualify and regulate cogenerators, are relatively new and are still being considered by the utility industry. Changes in government policy concerning other investment incentives, tariff, banking and wheeling facilities can also greatly affect the market.

2. Stricter air quality standards: Pollution regulations, could substantially add to the initial capital investment required for a cogeneration unit. The increased capital cost and in some cases the reduced system performance, will discourage the use of certain cogeneration technologies.

3. Changing utility buy-back rates: Electric utility buy-back rates are negotiated for each cogeneration site and are determined by the utility's avoided costs at that point in time. As the structure of the utility-industry changes and as more cogeneration capacity is brought on line, these buy-back rates may decrease and therefore, discourage further cogeneration.

4. Reluctance to enter the power business. Most manufacturing concerns have become accustomed to simply purchasing power and generating steam or process heat on-site. The prospect of adding a cogeneration system represents large capital investment in what is usually an unfamiliar technology.

5. Electric utility resistance: May electric utilities have adequate or excessive capacity at off-peak periods such as during nights and therefore have little incentive to purchase electrical power from a cogenerator. Other utilities disagree with the distributed on-site power concept, having grown up during a period when large central stations dominated the utility structure.

6. Lack of cogeneration expertise. Most industrial concerns do not have in-house personnel capable of evaluating, operating or maintaining cogeneration systems. In general, cogeneration technology expertise has been confined to specific industries, which have traditionally utilised equipment of this type.

7. Perceived unreliability of gas turbines: Gas turbines, which are typically perceived as high speed and sensitive equipment, are not generally utilised in many industrial prime mover applications. Most gas turbines are used in power producing applications as peaking generator sets and do require frequent maintenance due to the severe start-up and shutdown operating cycle.

8. Limited capital resources: Many industrial concerns do not have sufficient internal funds, or are unwilling or unable to obtain external funds.

9. Higher priority of production-oriented projects: Cogeneration projects are oriented towards energy savings and therefore take a lower priority than production-oriented projects. Higher returns on investment are typically required with energy saving projects when competing for limited capital budgets with production projects.

10. Sensitivity to energy prices: Small variations in the input and output utility costs of a cogeneration unit could easily destroy the economics of the project.

11. Possible plant closings: Companies will limit the scope of capital investment targeted at plants that may be severely curtailed or closed as a result of changing business or economic climates. Cogeneration projects typically require years of steady operation to be economically viable and therefore, are of limited interest in this situation.

12. Premium fuel availability: Of late, natural gas and oil are available at reasonable prices to the industrial sector throughout the year. The long-term availability of reasonably priced premium fuel is crucial to the successful implementation of several cogeneration technologies.

13. Limited floor area: Most cogeneration systems are retrofitted into existing manufacturing facilities with limited available floor space.

14. High capital cost for cogeneration systems: The available prime mover technologies limit the potential sizes of cogeneration systems to relatively large and expensive units. The high initial capital cost of cogeneration systems is a barrier for industries with limited financial strength.

15. Proper size and thermal-to-electric output match: Accommodating the application's energy steam requirements with that available from various technologies is instrumental to a successful project. Most cogeneration technologies have relatively fixed thermal-to-electric outputs and most applications have daily and/ or seasonal variations in utility requirements.

16. Adequate water supply: Many cogeneration technologies, which utilise heat recovery boilers, require a steady supply of high quality feedwater.

17. New production technologies that reduce energy use: New technologies, which could drastically reduce the utility requirements of current industrial processes, are emerging. This potential change in energy use or change in the composition of energy required could alter the viability of certain cogeneration technologies.

11.7.1   Technological constraints

Cogeneration under the power maximisation mode requires high-pressure and high-temperature system parameters. High-pressure boilers, turbines and accessories are not available in sufficient quantities indigenously. Turbines at high pressure are presently imported. There is also a considerable technological gap between the various advanced configurations and systems available in the developed countries.

11.7.2   Financial constraints

One of the major problems for a firm interested in cogeneration is the relative capital intensity of a cogeneration system. The capital cost of installation is very high.

Hence, the industries find difficult to raise resources for the incremental cost of setting up a cogeneration system with the objective of selling surplus power to the utilities.

11.7.3   Legislative constraints

At present, there is no legal framework for the purchase of surplus cogenerated power in India unlike in the U.S. In U.S. it is mandatory for public utilities to buy the energy generated by the cogenerator at the avoided cost under the Public Utility Regulations Policies Act (PURPA).

11.7.4   Technical barriers and problems of parallel operations

The quality of electric power has always been a major problem with the utilities in India. High voltage fluctuations, frequency variations and frequent power failures are factors that need to be taken care of. They add to the cost of the cogeneration plant.

Another consideration is the generator size rather than the plant load and the load on the utility life. This will usually result in more stringent protective requirements, since the utility and the cogenerator will again seek to protect utility and customer equipment from service disruptions.

11.8   Economic considerations

Steam turbine generators can help save substantial amount of energy in cogeneration systems. However, these plants require greater capital expenditures than plants which supply only low pressure process steam. This incremental capital investment must be justified by the reduced annual operating costs. Generally, a capital budgeting decision analysis is made to determine if the venture meets the company's requirement for return on investment or payout. Likewise, any additional investment for a larger or more efficient cogeneration alternative versus a smaller system must also meet management's requirements for return on investment or payout.

11.9   Risks and risk containment

A cogeneration is subject to a number of varied risks. Some of these are summarised below.

1. Design and technology: Will the proposed design and technology perform as anticipated at the appropriate efficiency levels?
2. Construction costs: Will the project be constructed within the estimated costs?
3. Timing and completion: Is the project subject to significant delays or will it be completed within the schedule as anticipated?
4. Fuel supply and cost: Will the required fuel supply be available in sufficient quantities and without significant escalation in cost?
5. Sale prices and quantities: Will there be an adequate market for the sale of cogeneration products and will the market accept the prices that are anticipated?
6. System performance: Will the various components of equipment in a cogeneration project perform as anticipated and produce the required outputs for the designated level of inputs?
7. Operating costs: Are the operating costs subject to unforeseen escalations?

8. Financing costs: Are the financing costs controlled, or is there a risk that these costs might escalate if economic conditions change?

9. Tax regulations: Will the anticipated tax benefits actually be available, or would changes in tax regulations remove some of these benefits?

10. Regulatory and environmental risks. Are there any significant changes in environmental or other regulations anticipated during the life of the project that may make the project unacceptable to regulatory authorities?

While all these risks cannot be totally eliminated, it is important to limit or contain these risks so that both the project developer and the financing organisation are comfortable with the level of risk faced by each, relative to the returns anticipated. Typical methods that have been used to contain risks are summarised below:

1. Design and technology risks: Obtain quality participants, who are knowledgeable regarding cogeneration technologies and design considerations.

2. Construction costs: Obtain insurance coverage against cost escalation, or assign fixed price contracts with the construction management firm.

3. Timing and completion risks: Obtain insurance protection or contract protection to assure that the project is completed on time.

4. Fuel supply and costs: Sign long-term contracts with fuel suppliers.

5. Electric and thermal sales: Obtain long-term contract protection by signing contracts with the purchasers of electricity and thermal energy.

6. System performance: Obtain performance guarantees from the manufacturer, or insurance coverage from appropriate insurance companies.

7. Operating costs: Obtain recognised and reputable design engineers, who are knowledgeable of the appropriate technologies being used.

8. Financing costs: Negotiate appropriate arrangements with the financing sources so that escalations of financing costs are minimised.

9. Tax regulations: Keep track of current trends in regulatory initiatives at both the central and state levels.

10. Environmental and other regulations: Keep track of the appropriate regulatory changes anticipated and obtain reputed organisations knowledgeable of environmental and other regulations relevant to the cogeneration project.

To sum up cogeneration is a useful technology in reducing waste and in conservation of resources. However, its application needs careful evaluation because of its complexity. In spite of the barriers and constraints, there is a need to enhance its penetration for the overall benefit of the society.

Energy efficient boilers

12.1   Introduction

The steam generator, or boiler, is a combination of systems and equipment for the purpose of converting chemical energy from fossil fuels into thermal energy and transferring the resulting thermal energy to a working fluid, usually water, for use in high-temperature processes or for partial conversion to mechanical energy in a turbine. In most modern large power plants, one boiler is used to supply steam to one steam-turbine generator unit. The boiler complex includes the air-handling equipment and ductwork, the fuel-handling system, the water-supply system, the steam drums and piping, the exhaust-gas system and the pollution-control system.

12.2   Boiler for textile industry

Textile industry is a very complex industry including spinning mills, weaving mills, knitting mills, dyeing mills, garments. Steam is widely used in textile industry. In every process of textile, in spinning, weaving, processing and garments steam is used for drying, heating and maintaining the temperature of system. Steam boiler in textile plants use fuels such as—LPG, natural gas, diesel oil, light oil, biomass fuel, rice husk, wood pellet, bagasse, palm kernel shell. Steam boilers provide sufficient steam for textile processing in drying, heating and maintaining the temperature of system.

Water tube steam biomass boiler in textile plant: Biomass is an ideal alternative fuel in boilers. The biomass water tube steam boiler can burn rice husk, straw, palm kernel shell, bagasse, wood pellet to produce steam for textile industry.

12.3   Functions of a boiler

A boiler has emerged as an important tool of industry with a high degree of versatility. As a source of hot water and/or steam it has found applications in a variety of industries like aluminium, automobiles, concrete block and bricks, ceramic, glass, inorganic and organic chemicals, copper primary and secondary, lumber, pulp and paper, selected plastics, rubber, textiles and sugar, etc. While a modern industrial boiler in the shape of a shell type multitubular package boiler is an example of optimised design and efficiency, no single design can obviously satisfy the need of such a wide range of requirements. The package boiler as we know it today combines in itself features of some of the earliest designs suitably modified to take care of changing demands. It may thus be useful to look at some of them keeping in mind the reasons that led to more and more improved versions culminating in the existing package boiler. Apart from the above mentioned use for process heat in industries, boiler has also been increasingly employed for power generation specially for captive power plants, where the demand are more severe.

The heat-transfer sections of a large boiler include the evaporator, the superheater, the re-heater, the air pre-heater and the economiser sections. The evaporator, superheater and re-heater surfaces are called primary heat-transfer surfaces while the air pre-heater and economiser surfaces are called secondary heat-transfer surfaces.

The heat-transfer sections and an energy flow diagram for a typical large pulverised-coal boiler are shown in Fig. 12.1.

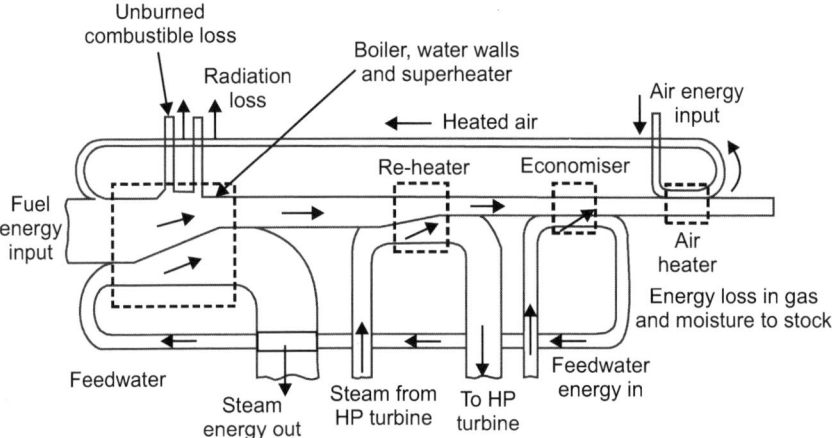

Figure 12.1: The layout and energy flow diagram for a typical pulverised-coal steam boiler.

12.4 Heating surfaces in a boiler

The amount of heating surface of a boiler is expressed in square meters. Any part of the boiler metal that actually contributes to making steam is a heating surface. The larger the heating surface a boiler has, the higher will be its capacity to raise steam.

Heating surfaces can be classified into several types:

1. Radiant heating surfaces (direct or primary) include all water-backed surfaces that are directly exposed to the radiant heat of the combustion flame.

2. Convection heating surfaces (indirect or secondary) include all those water-backed surfaces exposed only to hot combustion gases.
3. Extended heating surfaces include economisers and super heaters used in certain types of water tube boilers.

12.5 Boiler types and classification

Broadly, boilers can be classified into four types: fire tube boilers, water tube boilers, packaged boilers and fluidised bed combustion boilers.

12.5.1 Fire tube boilers

Fire tube boilers contain long steel tubes through which the hot gases from a furnace pass and around which the water to be converted to steam circulates. It is used for small steam capacities (up to 12000 kg/hr and 17.5 kg/cm^2). The advantages of fire tube boilers include their low capital cost and fuel efficiency (over 80%). They are easy to operate, accept wide load fluctuations and because they can handle large volumes of water, produce less variation in steam pressure. Flow diagram of fire tube boiler is shown in Fig. 12.2.

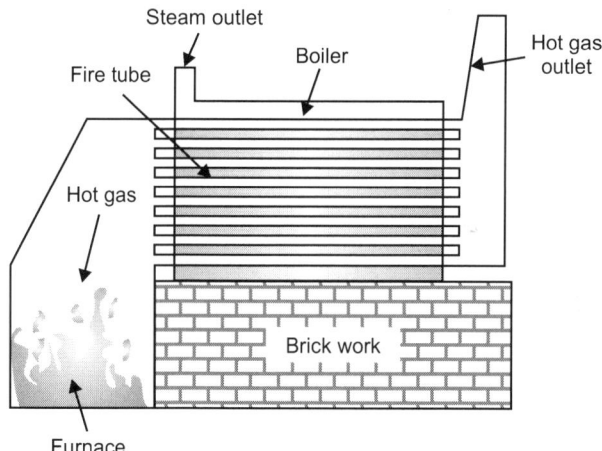

Figure 12.2: Fire tube boilers.

12.5.2 Water tube boilers

In water tube boilers, water passes through the tubes and the hot gasses pass outside the tubes. These boilers can be of single- or multiple-drum type. They can be built to handle larger steam capacities and higher pressures and have higher efficiencies than fire tube boilers. They are found in power plants whose steam capacities range from 4.5–120 T/hr and are characterised by high capital cost. These boilers are used when high pressure high-capacity steam production is demanded. They require more controls and very stringent water quality standards. Flow diagram of water tube boiler is shown in Fig. 12.3.

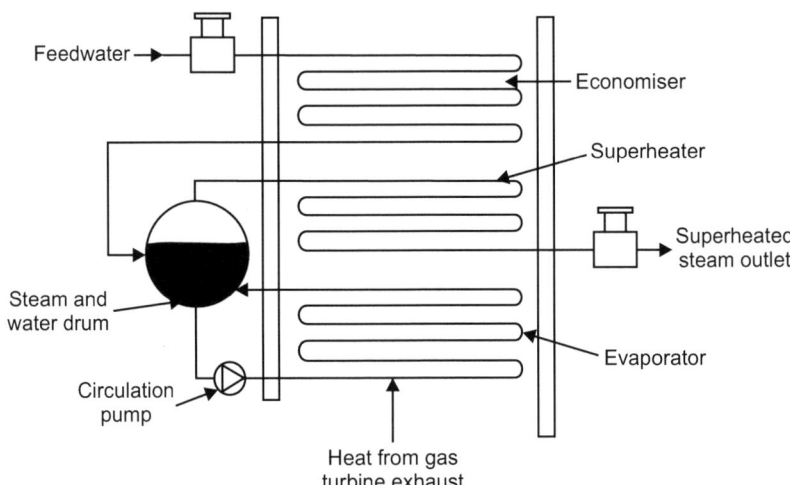

Figure 12.3: Water tube boilers.

All water-tube generators may be classified as either natural-circulation or forced-circulation boiler. Most of the older steam generators are natural-circulations systems. In these units, the fluid flow is induced by the difference in the specific weight of the water in the vertical supply tubes and the average specific weight of the water-steam mixture in the evaporator tubes where the evaporation occurs. The driving force for fluid flow through the evaporator section is equal to the product of the difference in the average specific weights and the height of the evaporator tubes. This driving force is balanced by the frictional pressure drop in the supply and evaporator tubes.

Most of the pressure drop occurs in the evaporator tubes as the result of the two-phase flow of water and steam. The principle of circulation in any natural-circulation boiler is shown in Fig. 12.4. The saturated water flows from the steam drum high in the boiler, through the supply or 'downcomer' tubes, located in the cooler part of the boiler, to the bottom or 'mud' drum. From the mud drum, the water flows back to the steam drum through the evaporator or 'riser' tubes. In the steam drum, the steam and water are separated and the steam is washed and dried before it is sent to be superheater. The internals of a typical steam drum are also presented in Fig. 12.4. The lower drum is commonly called the mud drum because any water impurities naturally accumulate in this drum. Periodically, some of the water in the mud drum is vented ('blow down') to the atmosphere and fresh water ('makeup') is added to lower the overall concentration of impurities in the water system.

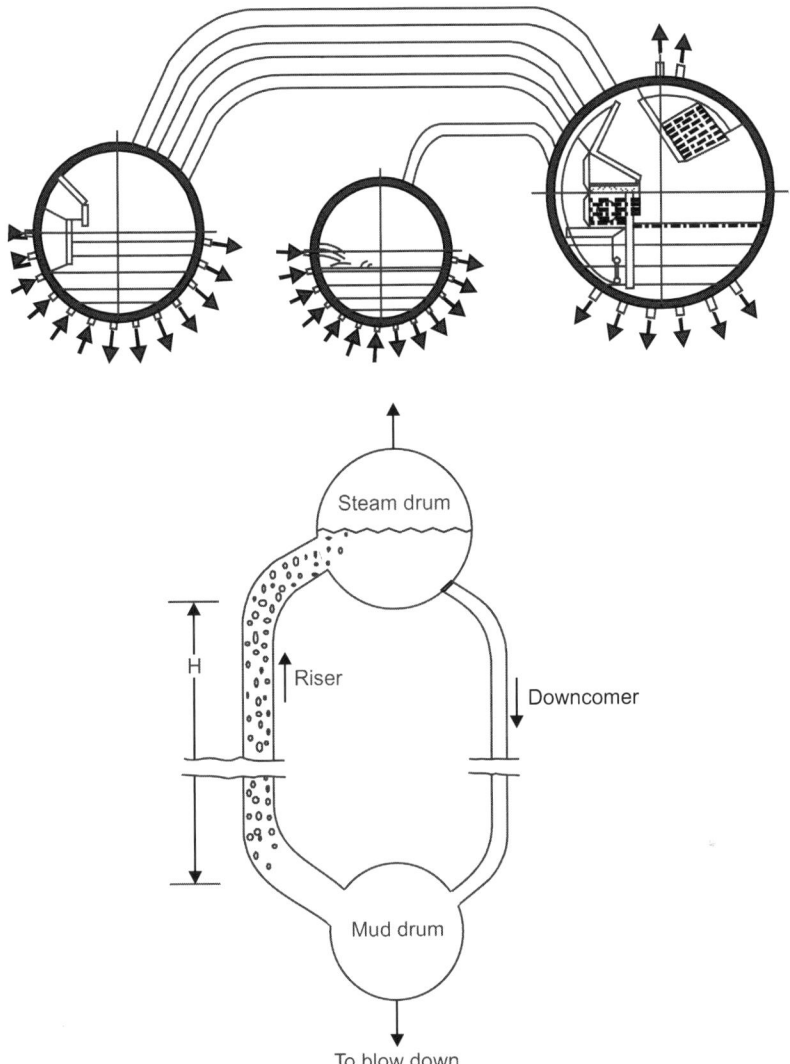

Figure 12.4: A typical steam drum and the schematic of a natural-circulation boiler.

As the steam pressure is increased, the density and specific gravity difference between the saturated steam and water decreases, requiring a greater height for adequate flow (up to 15 stories high). At the critical point of water (705.4°F and 3206.2 lb/in^2 abs) the difference is zero and there is no natural-circulation driving force. Actually, at pressures greater than about 160 atm (2400 lb/in^2 abs), the density difference is effectively too small for natural-circulation units and these systems must employ forced circulation.

12.5.3   Forced-circulation boiler

In a forced-circulation boiler, the fluid is pumped through the evaporator section of the boiler. This permits operation of the cycle at very high pressures, even above the critical pressure. High-pressure operation theoretically improves the efficiency of the basic steam cycle. The forced-circulation system eliminates the need for boiler height; it is lighter in weight; it uses smaller tubes and drums or no drums at all and the lower total water content in the boiler reduces the danger of a steam explosion. In addition to the problems associated with the main circulation pump and potentially higher operating pressure, some of these systems also require extremely pure water—orders of magnitude purer than that required for natural-circulation systems.

There are many different kinds of forced-circulation boilers, depending upon the circulation paths in the evaporator. The most widely used forced-circulation boilers system is the universal-pressure (UP) or Benson boiler. In the Benson boiler, the water is pumped to about 340 atm (5000 lb/in^2 abs) in the main feed pump (FP). The compressor water is then piped to the economiser section (E), through the evaporator tubes (T), through a transition section (TS) and finally through a convection superheater (CS), where it is exhausted to the turbine at a pressure of about 240 atm (3500 lb/in^2 abs). This supercritical boiler requires extremely pure water with an impurity concentration of only a few parts per billion. Any impurities at all in the water will normally be deposited in the boiler tubing. The Ramsin boiler is essentially identical to the Benson boiler except that the evaporating section is composed of inclined 'T'-bundle coils arranged in a spiral array.

12.5.4   Lancashire boiler

Lancashire is a horizontal, stationary fire tube boiler. This boiler was invented by Sir William Fairbairn in the year 1844. The flue gases flows through the fire tube, situated inside the boiler shell, so it is a fire tube boiler. This boiler generate low pressure steam. It a internally fired boiler because the furnace is placed inside the boiler.

This boiler works on the basic principle of heat exchanger. It is basically a shell and tube type heat exchanger in which the flue gases flow through the tubes and the water flows through shell. The heat is transfer from flue gases to the water through convection. It is a natural circulation boiler which uses natural current to flow the water inside the boiler.

Construction: As discussed above this boiler is similar to shell and tube type heat exchanger. It consists of a large drum of diameter up to 4–6 meter and length up to 9–10 meter. This drum consist two fire tube of diameter up to 40% of the diameter of shell. The water drum is placed over the bricks works.

Three spaces create between the drum and the bricks, one is at bottom and two are in sides as shown in Fig. 12.5. Flue gases passes through the fire tubes and side and bottom space. The water level inside the drum is always above the side channels of flue gases, so more heat transfer to the water. The drum is half filled with water and the upper half space for steam. The furnace is located at one end of the fire tubes inside the boiler. The low brick is situated at the grates (space where fuel burns) which does not allow to un-burned fuel and ash to flow in fire tubes. The boiler also consist other necessary mountings and accessories like economiser, super heater, safety valve, pressure gauge, water gauge, etc., to perform better.

Working of lancashire boiler

The Lancashire boiler is a shell and tube type heat exchanger. The fuel is burn at the grate. The water is pumped into the shell through the economiser which increases the temperature of water. Now the shell is half filled with water. The fire tube is fully immersed into the water. The fuel is charged at the grate which produces flue gases. These flue gases first passes through the fire tube from one end to another. This fire tubes transfer 80–90% of total heat to the water. The backward flue gases passes from the bottom passage where it transfer 8–10% heat to water.

The remaining flue gases passes from the side passage where it transfer 6–8% of heat to water. The brick is the lower conductor of heat, so work as heat insulator. The steam produces in drum shell and is taken out from the upper side where it flows through super heater if required. So the steam produce is taken by out for process work.

Advantage of lancashire boiler

1. This boiler is easy to clean and inspect.
2. It is more reliable and can generate large amount of steam.
3. It requires less maintenance.
4. This boiler is a natural circulation boiler so lower electricity consumption than other boilers.
5. It can easily operate.
6. It can easily meet with load requirement.
7. Lancashire boiler has high thermal efficiency about 80–90%.

Disadvantages of lancashire boiler

1. This boiler requires more floor space.
2. This boiler has leakage problem.
3. It requires more time to generate steam.

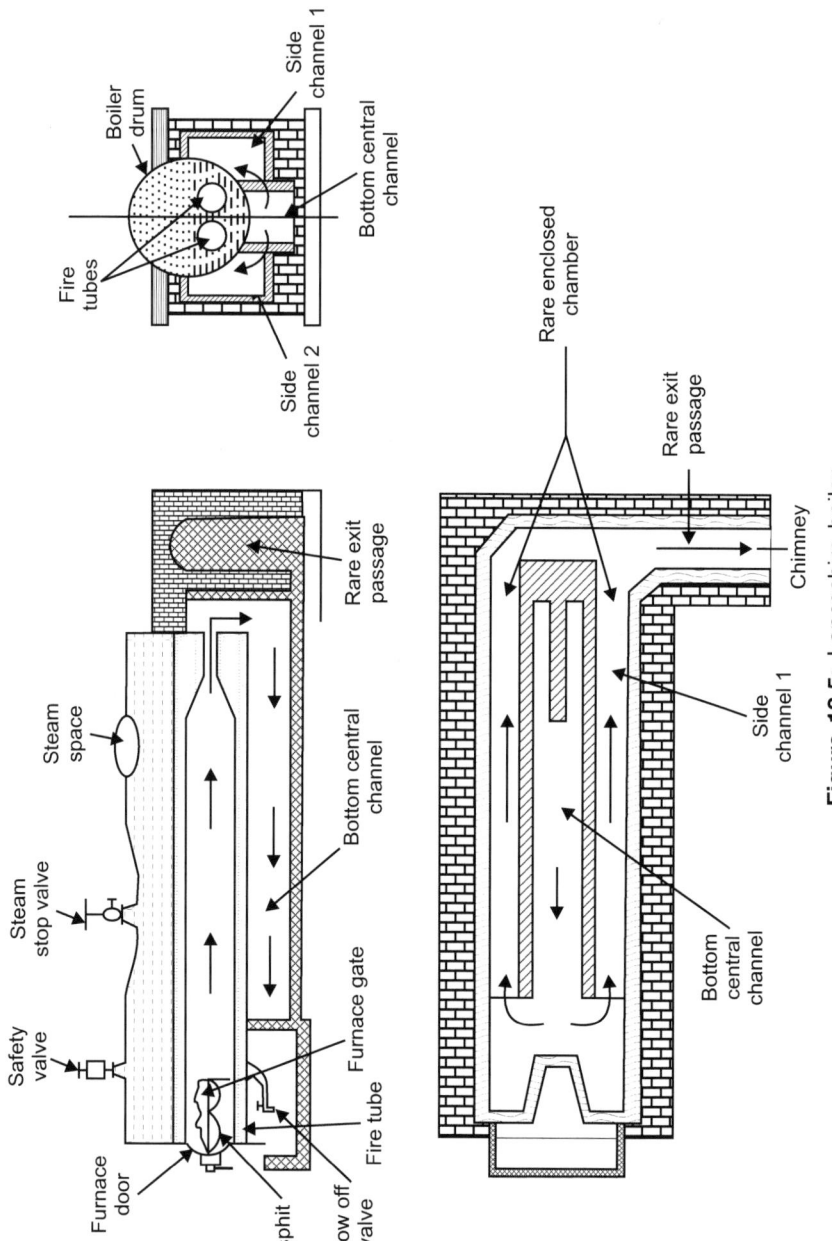

Figure 12.5: Lancashire boiler.

4. It cannot generate high pressure steam if required.

5. Grates are situated at the inlet of fire tube, which has small diameter. So the grate area is limited in this boiler.

12.5.5   Economic boiler

This is a horizontal fire tube boiler. It has no external flue ways, all gas passes within the boiler shell. It has therefore no problems connected with air leakage through external brick work settings. The diameter and length of its shell containing the main more fire tube range between 1.5 m to 2.7 m and 2.1 m to 6.7 m respectively. The fire tube runs from front to the back of the boiler, with the fire grate or any other firing mechanism being in the front part of the tube. The hot gases after reaching the end of the tube are reversed through tubes with a diameter of about 0.06 m to 0.08 m and surrounded by a refractory fire-brick lined chamber. In a two-phase boiler they may be made to pass to a smoke box located on the boiler front onto the main flue. In large size, called three-pass boilers, the gases on reaching the front of the boiler are passed a third time through the water space before finally going out of the rear end.

As in all shell boilers, the working pressure is limited to 15 to 20 bars and capacity ranges from about one-half to 10 T per hour. The heating surfaces within the boiler water space are disposed so as to give a much better heat transfer than the Lancashire boiler. This results in a quick raising of the steam head. Being compact with limited steam space, without a provision of a thermal storage, is not suited for widely fluctuating loads.

12.5.6   Multitubular package boiler

Packaged boilers

The packaged boiler is so called because it comes as a complete package. Once delivered to a site, it requires only steam, water pipe work, fuel supply and electrical connections in order to become operational. Package boilers are generally of shell type with fire tube design so as to achieve high heat transfer rates by both radiation and convection. These boilers are classified on the basis of the number of passes (the number of times the hot combustion gases pass through the boiler). The combustion chamber is taken as the first pass, after which there may be one, two, or three sets of fire tubes. The most common boiler of this class is a three-pass unit with two sets of fire tubes and with the exhaust gases exiting through the rear of the boiler.

Fluidised bed combustion (FBC) boilers

In fluidised bed boilers, fuel burning takes place on a floating (fluidised) bed in suspension. When an evenly distributed air or gas is passed upward through a finely divided bed of solid particles such as sand supported on a fine mesh, the particles are undisturbed at low velocity. As air velocity is gradually increased, a stage is reached when the individual particles are suspended in the air stream. A further increase in velocity gives rise to bubble formation, vigorous turbulence and rapid mixing and the bed is said to be fluidised. Fluidised bed boilers offer advantages of lower emissions, good efficiency and adaptability for use of low calorific-value fuels like biomass, municipal waste, etc.

12.5.7 Performance evaluation of boilers

The performance of a boiler, which include thermal efficiency and evaporation ratio (or steam to fuel ratio), deteriorates over time for reasons that include poor combustion, fouling of heat transfer area and inadequacies in operation and maintenance. Even for a new boiler, deteriorating fuel quality and water quality can result in poor boiler performance. Boiler efficiency tests help us to calculate deviations of boiler efficiency from the design value and identify areas for improvement.

12.6 Heat recovery and other appliances attached to the boiler

This section deals with waste heat boilers, economiser, air pre-heater and superheaters.

12.6.1 Waste heat boilers

As the name implies this type of boiler is based on the waste heat or the stack heat that may be available in any industry as the source of heat. Both the shell type and the water-tube type are usable for this purpose. The waste gases from the exhaust of the industrial plant are led via a ducting to the inlet chamber of the boiler. In a multi-tubular shell boiler they pass through the smoke tubes situated in the water space. It is necessary to increase the velocity of the gases passing through the tubes to achieve an effective heat transfer after breaking down the gas film that tends to adhere and stagnates on the tubes.

12.6.2 Economiser

An economiser is a heat recovery system designed to exploit the waste heat in the flue gases of a boiler. This consists in heating of the feedwater by the flue gases before it is fed to the boiler. In that sense it is at time considered as an extension of the boiler heating surface by which the evaporation capacity of the boiler is increased. It is normally rectangular in shape. Its length may be between 1 m to 10 m, while the breadth and heights will be determined by the capacity of the boiler. It consists of bank of tubes which lie in the path of the feedwater tank and the boiler. Thus the water flows through the tubes while the hot flue gases flow over their exterior. The tubes can be arranged either vertically or horizontally. The former type is amenable to easy cleaning externally and is preferred. The water is made to enter under pressure by a feed pump. The design of an economiser must take into account this fact while choosing the metal for its construction. Ordinary cast iron has been found to be a very suitable material for pressures up to about 20 bars for the vertical tube type. For strengthened form of vertical tube economiser the limitation of pressures may go up to 40 bars or so.

The pressure limit may be further extended by 15 bars in the case of the horizontal tube type. Cast iron is also well suited as a material from the point of view of corrosion resistance. As has been shown later there is a minimum temperature of the flue gases up to which they can be cooled. This limit is set by the dew points of water or for fuels containing sulphur by the dew point of the sulphuric acid. That gets formed by the sulphur trioxide dissolving in the water contained in the flue gases. Acceptable temperature for cast iron are 100–120°C.

12.6.3   Air pre-heater

Air pre-heater is still another heat recovery appliance that also leans on the heat contained in the flue gases. This could be used both as an independent unit or in conjuction with an economiser. This is based on pre-heating the primary air required for combustion. This is achieved through the use of an air-to-air heat exchanger. There are several advantages of using pre-heated air for combustion. Some of them are higher furnace temperature and hence better heat transfer by radiation, instantaneous combustion of the inflammable volatiles in the fuel. Pre-heated air also gives rise to short hot flame when the fuel is bituminous coal thereby reducing smoke. Last but not the least, a low graded coal can be burnt with low excess air.

The air pre-heaters may be designed either on the regenerative or recuperative principle. The latter has a configuration similar to an economiser, in which the cold outside air flows inside the tubes while the hot flue gases flow outside. Unlike an economiser it is not a high pressure unit. Cast iron is a preferred material even for this in view of its anti-corrosion like properties.

12.6.4   Superheaters

The function of a superheater is to completely dry the almost dry steam taken from a boiler and to heat it further at almost constant pressure. Superheater, by and large, form and integral part of the boiler. In the fire-tube boiler, the superheater is placed just after the first pass of the hot flue gases through the boiler. At this stage most of the heat transfer takes place as the temperature of the gases is high enough. The conductivity of the vapour inside the superheater being poor, the tube material of the superheater has to be different from that of the boiler to be able to withstand much higher temperatures. A typical boiler with a superheater and an economiser is shown in Fig. 12.6.

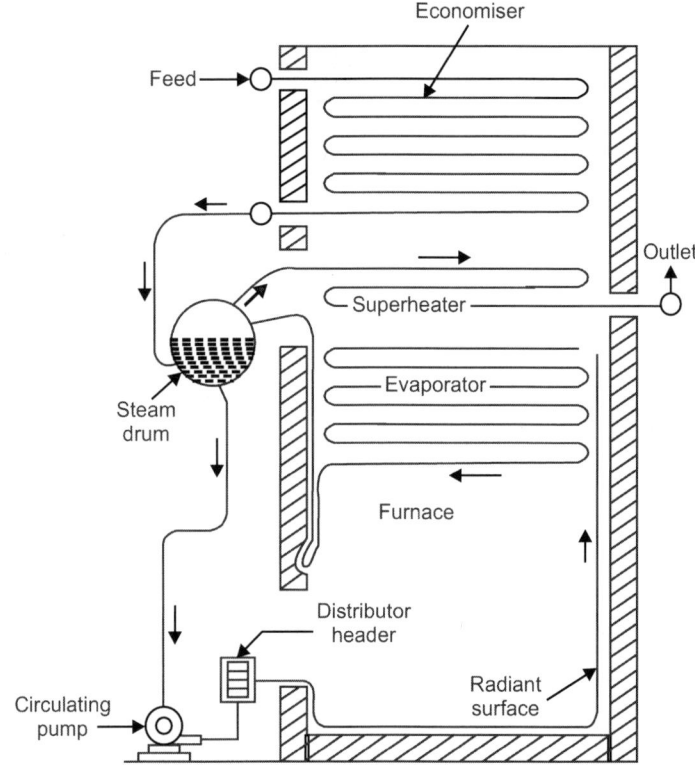

Figure 12.6: Forced circulation boiler with economiser and superheater.

12.6.5   Some other boiler auxiliaries

A boiler suitably mounted has to have a few auxiliaries before it may be considered fully functional. These broadly consists of boiler feed auxiliaries and mountings that form an integral part of the boiler. Under the former comes a feedwater pump and a tank and regulators. The pumps may either be reciprocating or rotary types driven by electricity or by steam. The regulators are meant to automatically control the rate of feeding water into the boiler so as to offset the surges of large quantities of water. This helps ensure steadier combustion rate and saving of fuel.

The mountings are safety and control mountings. The former includes safety valves, water gauges, fusible plugs and high and low water alarms. The control mountings consist of pressure gauges, feedwater check valves, blow down valves, the main stop valve and the dampers.

Apart from the above boiler has a control system that consists of a steam pressure transducer responding to load demand. The corresponding fuel requirement generated is controlled by a full flow controller. The matching need of air for the varying fuel flow rates is met through either a mechanical linkage or a pneumatic transducer connecting the fuel valve with the air damper. It is practically impossible with this kind of mechanical arrangement to neutralise rapidly fluctuating claims on air resulting from changes in fuel or air quality. It is thus customary in such systems to operate with damper settings at more excess oxygen than necessary to take care of transitory aberrations.

12.7   Thermal efficiency

Thermal efficiency of a boiler is defined as the percentage of heat input that is effectively utilised to generate steam. There are two methods of assessing boiler efficiency: direct and indirect. In the direct method, the ratio of heat output (heat gain by water to become steam) to heat input (energy content of fuel) is calculated. In the indirect method, all the heat losses of a boiler are measured and its efficiency computed by subtracting the losses from the maximum of 100.

12.7.1   Evaporation ratio

Evaporation ratio, or steam to fuel ratio, is another simple, conventional parameter to track performance of boilers on-day-to-day basis. For small capacity boilers, direct method can be attempted, but it is preferable to conduct indirect efficiency evaluation, since an indirect method permits assessment of all losses and can be a tool for loss minimisation. In the direct method, steam quality measurement poses uncertainties.

Standards can be referred to for computations and methodology of evaluation.

Example of direct efficiency calculation:
Calculate the efficiency of the boiler from the following data:

Type of boiler	:	Coal-fired
Quantity of steam (dry) generated	:	8 TPH
Steam pressure (gauge)/temp	:	10 kg/cm^2 (g)/180°C
Quantity of coal consumed	:	1.8 TPH

Feed water temperature	:	85°C
GCV of coal	:	3200 kcal/kg
Enthalpy of steam at 10 kg/cm² (g) pressure	:	665 kcal/kg (saturated)
Enthalpy of inlet fed water	:	85 kcal/kg

$$\text{Bioler efficiency } (\eta) = \frac{8 \text{ TPH} \times 1000 \text{ kg} \times (665 - 85) \times 100}{1.8 \text{ TPH} \times 1000 \text{ kg} \times 3200}$$

$$= 80.0\%$$

Evaporation ratio = 8 TPH of steam/1.8 TPH of coal = 4.4

12.8   Environmental compliance of boilers

Environmental aspects are becoming more important day by day. Generally local pollution control boards have limits specified for polluting elements in flue gases. Constituents such as CO, NO_x, particulate matters, SO_2/SO_3, hydrocarbons, in flue gases should be measured and must be below the limits specified by the pollution control boards.

Following aspects are equally important even though many of them do not get covered under any statutory requirements:

1. Boiler water TDS and treatment before disposal.
2. No fuel oil spillage/proper spillage recovery (for oil fired boilers).
3. Noise level in boiler house.
4. Normal boiler house ambient.
5. Proper soot disposal system while tube cleaning and after tube cleaning.
6. Ash disposal system (for solid fuel boilers).
7. Lighting and illumination in boiler house.
8. Fire extinguishers in boiler house.
9. First aid kit in boiler house.
10. Space for operator movement.

12.9   Energy conservation opportunities in boilers

The various energy efficiency opportunities in boiler system can be related to combustion, heat transfer, avoidable losses, high auxiliary power consumption, water quality and blow down. Examining the following factors can indicate if a boiler is being run to maximise its efficiency:

Stack temperature: The stack temperature should be as low as possible. However, it should not be so low that water vapour in the exhaust condenses on the stack walls. This is important in fuels containing significant sulphur as low temperature can lead to sulphur dew point corrosion. Stack temperatures greater than 200°C indicates potential for recovery of waste heat. It also indicate the scaling of heat transfer/recovery equipment and hence the urgency of taking an early shut down for water/flue side cleaning.

Feed water pre-heating using economiser: Typically, the flue gases leaving a modern 3-pass shell boiler are at temperatures of 200 to 300°C. Thus, there is a potential to recover heat from these gases. The flue gas exit temperature from a boiler is usually maintained at a minimum of 200°C, so that the sulphur oxides in the flue gas do not condense and cause corrosion in heat transfer surfaces. When a clean fuel such as natural gas, LPG or gas oil is used, the economy of heat recovery must be worked out, as the flue gas temperature may be well below 200°C. The potential for energy saving depends on the type of boiler installed and the fuel used. For a typically older model shell boiler, with a flue gas exit temperature of 260°C, an economiser could be used to reduce it to 200°C, increasing the feed water temperature by 15°C.

Increase in overall thermal efficiency would be in the order of 3%. For a modern 3-pass shell boiler firing natural gas with a flue gas exit temperature of 140°C a condensing economiser would reduce the exit temperature to 65°C increasing thermal efficiency by 5%.

Combustion air pre-heat: Combustion air pre-heating is an alternative to feed water heating. In order to improve thermal efficiency by 1%, the combustion air temperature must be raised by 20°C. Most gas and oil burners used in a boiler plant are not designed for high air pre-heat temperatures. Modern burners can withstand much higher combustion air pre-heat, so it is possible to consider such units as heat exchangers in the exit flue as an alternative to an economiser, when either space or a high feed water return temperature make it viable.

Incomplete combustion: Incomplete combustion can arise from a shortage of air or surplus of fuel or poor distribution of fuel. It is usually obvious from the colour or smoke and must be corrected immediately.

In the case of oil and gas fired systems, CO or smoke (for oil fired systems only) with normal or high excess air indicates burner system problems. A more frequent cause of incomplete combustion is the poor mixing of fuel and air at the burner. Poor oil fires can result from improper viscosity, worn tips, carbonisation on tips and deterioration of diffusers or spinner plates.

With coal firing, unburned carbon can comprise a big loss. It occurs as grit carry-over or carbon-in-ash and may amount to more than 2% of the heat supplied to the boiler. Non uniform fuel size could be one of the reasons for incomplete combustion. In chain grate stokers, large lumps will not burn out completely, while small pieces and fines may block the air passage, thus causing poor air distribution. In sprinkler stokers, stoker grate condition, fuel distributors, wind box air regulation and over-fire systems can affect carbon loss. Increase in the fines in pulverised coal also increases carbon loss.

Excess air control: Table 12.1 gives the theoretical amount of air required for combustion of various types of fuel. Excess air is required in all practical cases to ensure complete combustion, to allow for the normal variations in combustion and to ensure satisfactory stack conditions for some fuels. The optimum excess air level for maximum boiler efficiency occurs when the sum of the losses due to incomplete combustion and loss due to heat in flue gases is minimum.

Table 12.1: Theoretical combustion data–common boiler fuels.

Fuel	*Kg of air req./kg of fuel*	*Kg of flue gas/kg of fuel*	*m^3 of flue/kg of fuel*	*Theoretical CO_2% in dry flue gas*	*CO_2% in flue gas achieved in practice*
Solid fuels					
Bagasse	3.2	3.43	2.61	20.65	10–12
Coal (bituminous)	10.8	11.7	9.4	18.7	10–13
Lignite	8.4	9.1	6.97	19.4	9–13
Paddy Husk	4.6	5.63	4.58	19.8	14–15
Wood	5.8	6.4	4.79	20.3	11.13
Liquid fuels					
Furnace Oil	13.9	14.3	11.5	15	9–14
LSHS	14.04	14.63	10.79	15.5	9–14

This level varies with furnace design, type of burner, fuel and process variables. It can be determined by conducting tests with different air fuel ratios. Typical values of excess air supplied for various fuels are given in Table 12.2.

Table 12.2: Excess air levels for different fuels.

Fuel	*Type of furnace or burners*	*Excess air (% by wt)*
Pulverised coal	Completely water-cooled furnace for slag-tap or dry-ash removal	15–20
	Partially water-cooled furnace for dry-ash removal	15–40
Coal	Spreader stoker	30–60
	Water-cooler vibrating-grate stokers	30–60
	Chain-grate and travelling-gate stokers	15–50
	Underfeed stoker	20–50
Fuel oil	Oil burners, register type	15–20
	Multi-fuel burners and flat-flame	20–30

(Cont'd...)

Fuel	Type of furnace or burners	Excess air (% by wt)
Natural gas	High pressure burner	5–7
Wood	Dutch over (10–23% through grates) and Hofft type	20–25
Bagasse	All furnaces	25–35
Black liquor	Recovery furnaces for draft and soda-pulping processes	30–40

Controlling excess air to an optimum level always results in reduction in flue gas losses; for every 1% reduction in excess air there is approximately 0.6% rise in efficiency.

Various methods are available to control the excess air:

1. Portable oxygen analysers and draft gauges can be used to make periodic readings to guide the operator to manually adjust the flow of air for optimum operation. Excess air reduction up to 20% is feasible.
2. The most common method is the continuous oxygen analyser with a local readout mounted draft gauge, by which the operator can adjust air flow. A further reduction of 10–15% can be achieved over the previous system.
3. The same continuous oxygen analyser can have a remote controlled pneumatic damper positioner, by which the readouts are available in a control room. This enables an operator to remotely control a number of firing systems simultaneously.

The most sophisticated system is the automatic stack damper control, whose cost is really justified only for large systems.

Radiation and convection heat loss: The external surfaces of a shell boiler are hotter than the surroundings. The surfaces thus lose heat to the surroundings depending on the surface area and the difference in temperature between the surface and the surroundings. The heat loss from the boiler shell is normally a fixed energy loss, irrespective of the boiler output. With modern boiler designs, this may represent only 1.5% on the gross calorific value at full rating, but will increase to around 6%, if the boiler operates at only 25 % output. Repairing or augmenting insulation can reduce heat loss through boiler walls and piping.

Automatic blow down control: Uncontrolled continuous blow down is very wasteful. Automatic blow down controls can be installed that sense and respond to boiler water conductivity and pH. A 10% blow down in a 15 kg/cm^2 boiler results in 3% efficiency loss.

Reduction of scaling and soot losses: In oil and coal-fired boilers, soot buildup on tubes acts as an insulator against heat transfer. Any such deposits should be removed on a regular basis. Elevated stack temperatures may indicate excessive soot buildup. Also same result will occur due to scaling on the water side. High exit gas temperatures at normal excess air indicate poor heat transfer performance. This condition can result from a gradual build-up of gas-side or waterside deposits. Waterside deposits require a review of water treatment procedures and tube cleaning to remove deposits. An estimated 1% efficiency loss occurs with every 22°C increase in stack temperature. Stack temperature should be checked and recorded regularly as an indicator of soot deposits. When the flue gas temperature rises about 20°C above the temperature for a newly cleaned boiler, it is time to remove the soot deposits. It is, therefore, recommended to install a dial type thermometer at the base of the stack to monitor the exhaust flue gas temperature. It is estimated that 3 mm of soot can cause an increase in fuel consumption by 2.5% due to increased flue gas temperatures. Periodic off-line cleaning of radiant furnace surfaces, boiler tube banks, economisers and air heaters may be necessary to remove stubborn deposits.

Reduction of boiler steam pressure: This is an effective means of reducing fuel consumption, if permissible, by as much as 1 to 2%. Lower steam pressure gives a lower saturated steam temperature and without stack heat recovery, a similar reduction in the temperature of the flue gas temperature results. Steam is generated at pressures normally dictated by the highest pressure/temperature requirements for a particular process. In some cases, the process does not operate all the time and there are periods when the boiler pressure could be reduced. The energy manager should consider pressure reduction carefully, before recommending it. Adverse effects, such as an increase in water carryover from the boiler owing to pressure reduction, may negate any potential saving. Pressure should be reduced in stages and no more than a 20% reduction should be considered.

Variable speed control for fans, blowers and pumps: Variable speed control is an important means of achieving energy savings. Generally, combustion air control is effected by throttling dampers fitted at forced and induced draft fans. Though dampers are simple means of control, they lack accuracy, giving poor control characteristics at the top and bottom of the operating range. In general, if the load characteristic of the boiler is variable, the possibility of replacing the dampers by a VSD should be evaluated.

Effect of boiler loading on efficiency: The maximum efficiency of the boiler does not occur at full load, but at about two-thirds of the full load. If the load on the boiler decreases further, efficiency also tends to decrease. At zero output, the efficiency of the boiler is zero and any fuel fired is used only to supply the losses.

The factors affecting boiler efficiency are:

1. As the load falls, so does the value of the mass flow rate of the flue gases through the tubes. This reduction in flow rate for the same heat transfer area, reduced the exit flue gas temperatures by a small extent, reducing the sensible heat loss.

2. Below half load, most combustion appliances need more excess air to burn the fuel completely. This increases the sensible heat loss.

In general, efficiency of the boiler reduces significantly below 25% of the rated load and as far as possible, operation of boilers below this level should be avoided

Proper boiler scheduling: Since, the optimum efficiency of boilers occurs at 65–85% of full load, it is usually more efficient, on the whole, to operate a fewer number of boilers at higher loads, than to operate a large number at low loads.

Boiler replacement: The potential savings from replacing a boiler depend on the anticipated change in overall efficiency. A change in a boiler can be financially attractive if the existing boiler is:

1. Old and inefficient.
2. Not capable of firing cheaper substitution fuel.
3. Over or undersized for present requirements.
4. Not designed for ideal loading conditions.

The feasibility study should examine all implications of long-term fuel availability and company growth plans. All financial and engineering factors should be considered. Since boiler plants traditionally have a useful life of well over 25 years, replacement must be carefully studied.

12.10   Improving boiler efficiency

Boilers are an integral utility of any process industry and are known to consume large amounts of fuel and electrical energy. With fluctuating oil prices impacting production costs, there has been an increasing awareness for upgradation, automation and efficiency improvement in boilers and to maintain the boilers in good health. In most industries pricing of the end product is mainly decided by fuel cost and energy bill. Among utilities, boilers and heating systems consume large amount of fuel and electrical energy. Since boilers come under IBR rules and operate with various fuels, they are considered stationary equipment and once you install them, it is difficult to replace or modify.

Although boiler technology has not seen any drastic changes in recent years, new avenues have opened up for upgradation, automation and efficiency improvement in boilers and heaters. It is a tough challenge for maintenance and engineering teams to constantly update newer methods and systems of boiler operations and in the process ensure efficient steam management.

One of the reasons for improvement of boiler operations over the last decade is the growing awareness of global warming and the need to implement stringent pollution control norms to reduce industrial emissions and arrest green house effect. As a result, though useful boiler life is said to be anywhere between 10 to 20 years, it is important to maintain boiler operations as efficient and healthy as in the case of new equipment.

Another factor that is forcing industry to go for cost effective solutions is the availability and environmental viability of fossil fuels. Shift to biomass fuels and use of multiple fuels are emerging as viable options.

Various avenues for industries to improve boiler efficiency and reduce down time include:

1. Correct use of fuel and combustion equipment.
2. Retrofits and upgrades.
3. Energy recovery.
4. Steam distribution.
5. Minimising break down in boilers and avoiding production loss.

12.10.1 Correct use of fuel and combustion equipment

It is important to use specified fuel for which the boiler is designed. If the boiler is designed for Indian coal, it cannot burn Indonesian or South African coal effectively. This is mainly due to the different fuel characteristics they possess. Indian coal is typically burnt in fluidised over bed combustion more effectively as it has low fines percentage and, high ash content, where as the imported coal burns more effectively when it is fired in fluidised under bed option. Neglecting such critical choice of equipment to match fuel requirements can lead to huge carry over, lower combustion efficiency, high operational and maintenance cost.

Similarly a boiler running on LDO/HSD, when changed to furnace oil (FO), it cannot work effectively without making necessary changes in the combustion system. Fuel characteristics differ widely for HSD, LDO and FO. It is a known fact that FO is difficult to atomise (creation of smaller particle for better mixing of fuel with combustion air), as it contains chain and ignition properties. Moreover, compared to other lighter fuels, it contains undesirable material like conradson carbon, asphaltines, sediments, silica, vanadium, etc., in excess. This is mainly due to depleting crude oil reserves pushing oil industries to extract maximum lighter hydrocarbons and FO is the final product that remains as a by-product.

For optimum efficiency, it is important to choose the right combustion equipment for specific fuel. Due to scarcity of specific fuels, users are forced to use unspecified fuels in boilers to generate steam. The result can be any one of the following issues:

1. Low combustion efficiency.
2. High fuel combustion per ton of steam generated.
3. Higher fouling of heat transfer area and increased stack losses.

4. Poor steam generation.

5. High maintenance cost of boiler components like nozzles, fuel pumps, dust controllers.

6. Corrosion of chimney and ducts.

7. Corrosion/erosion of ID fans, ducts, boiler tubes.

8. Higher emission of pollutants such as CO, NO_x, SO_x, etc.

9. Frequent stop and start of boilers and associated components.

10. Operational safety hazards.

Proper planning of fuel selection while installing a new boiler can help avoid the issues mentioned above at later stages. Sufficient space around boilers helps to implement future upgrades and modifications in case of changes in fuel or norms or if efficiency improvement is planned.

12.10.2 Retrofits and upgrades of boilers

It has become a common practice to retrofit, upgrade old boilers with new combustor and burner designs. This is healthier and more profitable than continuing with inefficient and unsafe equipment. A bagasse fired stationary grate boiler can be converted to operate on both bagasse as well as coal with a separate fluidised bed combustor added for higher thermal efficiency. This will give the boiler the flexibility of using one fuel while keeping a second fuel option.

Conventionally, industry uses pressure jet burners for most liquid oils. If a process demands trouble free operation of boilers with consistent efficiency and minimum possible stoppages for maintenance air/steam, atomised burners can be the best option. They also allow users to have multi-fuel options, where there is simultaneous burning of more than two fuels, something practically difficult in pressure jet burners.

A well designed steam/air atomised burner can help maintain committed boiler efficiency consistently for longer durations of 3–4 months. In pressure jet burners, combustion efficiency drops over time due to nozzle clogging, fuel pressure variations, etc. If proper maintenance on a weekly basis is not followed for burner components, a drastic drop of efficiency ranging from 1–2% is observed.

Today it is possible to step up the overall efficiency of boilers with new facets of combustion technology like oxygen trimming, oxygen control, variable frequency drive based air and draft control, boiler performance monitoring and measurement systems based on latest PLC/SCADA systems. A 20 degree rise in stack temperature can mean 1% efficiency loss and 1% oxygen saving in flue gas means 0.5% increase in thermal efficiency.

12.10.3 Energy recovery from boilers

In most boilers thermal efficiency is calculated as per BS 845 part 1 guidelines. It is clear from these guidelines that dry flue gas loss (stack loss) contributes 11 to 12% energy loss in conventional oil and gas boilers that don't have heat recovery. In solid fuel boilers these losses can be as high as 15 to 25% depending on the type of fuel, combustor, etc. A properly designed heat recovery unit at the flue gas exit point of a boiler/heater can save fuel cost and improve overall equipment efficiency. Some of the typical cost saving heat recovery options for boiler and heaters are:

1. Air pre-heater (integral/external)-saves fuel cost upto 4%.
2. Water pre-heater-saves fuel cost upto 3%.
3. IBR economiser-saves fuel cost upto 4%.
4. Condensing economiser-saves fuel cost upto 13% on natural gas fired boilers and heaters.

Similarly, good amount of energy can be recovered if users implement condensate recovery systems in the process plants.

A typical condensate recovery system comprising flash vessel, mechanical steam operated pump and de-aerator head can save up to Rs. 1 crore in fuel costs for a 10 TPH boiler running on FO.

Condensate recovery has the following major benefits:

1. It saves precious water used to generate steam as condensate water is one of the best sources of feed water for boiler.
2. As condensate is at a higher temperature it saves fuel cost, as feed water is already pre-heated upto 90°C.
3. A well designed mechanical pump saves expensive electrical cost of condensate transfer from process to utility.
4. Elevated temperature of condensate helps to remove dissolved oxygen in the feed water. This minimises corrosion of boiler internals, improves safety and extends boiler life.
5. Lesser use of fuel helps to reduce carbon dioxide generation and global warming.
6. Reduces chemical load on effluent treatment plants, resulting in reduced overall cost of operations and water pollution.
7. Pure condensate water reduces blow down loss up to 60%, further improving overall efficiency of boiler.

12.10.4 Steam distribution from boilers

Another aspect of energy saving comes from effective distribution and utilisation of steam from boiler to process.

While installing a steam distribution system, it is important to consider the following:

1. Optimum steam line sizing to avoid higher pressure drops, minimum heat loss, reduced water logging and hammering.
2. Properly sized and effective steam/condensate line insulation to minimise radiation and convection losses.
3. Air venting and vacuum breaker at proper locations to improve heat transfer from available steam.
4. Use of drain traps at recommended positions in steam lines (at least one trap at 30 meter straight distance) to remove condensate from steam lines to reduce water logging and hammering.
5. Correct selection of drain traps to minimise the steam loss—a thermodynamic trap for saturated steam line and thermostatic traps for steam tracing lines.
6. Use of right size and type of float traps for removal of condensate in indirect heating applications.

Fuel additives/water side chemicals like anti-descalents for heat transfer and fuel combustion also helps to maintain boiler efficiencies.

Minimising breakdown in boilers and avoiding production loss

There could be many reasons offered for boiler breakdowns and shutdowns. But it is better to realise the practical good sense in the observation what you can monitor you can correct and improve. In today's demanding workplace, maintenance managers and supervisors are under pressure to perform and keep the utilities in good working condition. Not every plant has the luxury of stand by boilers and it becomes the responsibility of utility managers to ensure minimum shutdowns and breakdowns. Although standard maintenance practices are helpful to a great extent to minimise breakdowns, with increasing complexity of utilities and demanding production schedules, an effective and automated maintenance schedule can be of real help.

Some of the newer concepts that can help utility managers are:

1. Install remote monitoring, where a computer system with simple instrumentation monitor the health of boilers and generate reports at pre-defined intervals by email or mobile phone messages (sms).
2. Plan a fixed preventive maintenance schedule for boiler, synchronised with production schedule.
3. Instead of hard copies maintain electronic log books that will generate analysis of various parameters.
4. Observe MTBF (mean time between failure) of various components of boiler and take correct action with OEMs and experts.

5. Prepare FTA (Fault Tree Analysis) for frequently occurring and critical problems to generate data base for everyone to follow. This can help new recruits and speed up resolution of problems.

6. Follow OEM instructions and best practices of maintenance.

7. Install boiler efficiency/performance monitoring system so that performance is system based and not person-dependent.

8. Do a Remaining Life Analysis (RLA) for pressure parts of boiler to predict effective safe life of boilers. This will also help to avoid safety related failures and hazards that can result in longer time to restore boilers.

9. Conduct energy/efficiency audits of boilers for fuel-energy savings and implement recommendations.

10. Use genuine parts and adopt good engineering practices while carrying out maintenance of boilers.

11. Train the operating teams at regular intervals by experts or OEMs to update knowledge. Reward people who imbibe and practice innovative ways of operating boilers.

12. Stock necessary critical parts to avoid production loss.

13. Do not bypass safety rules as it may lead to major safety hazards and breakdowns.

The performance of boilers ultimately depends on how well they are maintained and operated by trained professionals. Boiler operations in an industrial plant will be efficient and safe with a utility team that keeps track of healthy maintenance practices and implement them year after year.

12.11   Factors affecting boiler efficiency

There are many factors that affect boiler efficiency. It is important to adjust the boiler regularly for these variations in order to ensure optimum boiler performance.

12.11.1 Shift variations in boilers

Typically, in plants, the load on the plant changes from shift to shift. If the boiler response to these load variations is not changed, the efficiency of the boiler drops. The change in settings required can be as simple as changing the firing rate or pressure limits of the boiler. These can be done easily from the boiler control panel itself. These changes are easy to do in either oil, gas or solid fuel fired boilers. A change may also be required in feed water tank level settings as the load may be lower in certain shifts. Also the blow down rate needs to be adjusted to reduce blow down based on TDS and not based on time as the load is low. The combustion system also needs some fine tuning as the day time and night time temperatures may change by 15 to 20°C. This variation would affect the excess air setting of the burner and hence dampers may need to be adjusted.

12.11.2 Daily variations in boilers

Load variations may occur from day to day too. Again, the boiler needs to be adjusted to cater to this efficiently.

12.11.3 Weekly variations in boilers

As the fuel quality is changing continuously, adjustments need to be made to the combustion system to take care of these variations. In oil or gas fired boilers, the quality of oil being received may vary from tanker to tanker. The moisture content in solid fuels changes with time and also the source of purchase. This change in fuel quality makes adjustments necessary.

12.11.4 Seasonal variations in boilers

Boiler loading pattern is an important factor here too. The production requirement of the plant may be affected by seasonal demands. This calls for adjustments again. In solid fuel fired boilers, the fuel available may change depending on seasons. The ambient air temperatures and humidity will also change from season to season. The combustion system needs adjustment too.

12.12 Practical standard operating practices for improving boiler efficiency

Step 1: Steam loading (setting the pressure control loop right)

1. Boiler loading plays an important role in varying boiler efficiency.
2. The loading pattern can be well defined for most processes.
3. Setting of boiler control loops as per requirement.
4. Critical parameters - steam flow and pressure (temperature).
5. There is no other way to do this.
6. Losses - On/Off, combustion, radiation (in multiple boilers).

Step 2: Combustion losses (setting the air to fuel ratio right)

1. Having done step one, this can be done fairly easily.
2. Not only this, the first step holdown alignment support (HAS) to be done before attempting this for any meaningful output.
3. Critical parameters-stack oxygen, temperature and steam temperature.
4. Losses–combustion.

Step 3: Blow down (ABCO)

1. Gets affected by-FW TDS, CRR, TDS set point, Boiler loading.
2. Monitoring as a loss with others.
3. Critical parameters - TDS, FW temperature.
4. Losses-blow down.

12.12.1 Boiler efficiency improvement ladder-bridging the gaps

Gap 1: Direct efficiency of 72% and indirect efficiency of 78%

1. Losses like the chimney draft loss during standby, loss due to cold air purging during start up cycles and during standby and loss due to the on/off cycles of the burner are generally not measured and are very small in magnitude if the boiler is operated continuously. These factors play an important role if the boiler load is low or varying a lot.
2. By monitoring the steam flow patterns over a period of time, the peak and low load demands can be easily mapped out.
3. All the boiler efficiency products generate data critical for reducing fuel consumption and also keep a check on the process itself.

Gap 2: Average indirect efficiency of 78% and the best indirect efficiency of 84%

1. Improper tuning of burner.
2. This gap can be bridged easily by monitoring the stack oxygen and temperature on a regular basis.
3. The boiler loading and variations in fuel firing rate also play an important role in burner combustion. The burner tuning needs regular adjustments because of variation in fuel quality and burner nozzle condition too.
4. Bridging this gap does not mean investing in expensive control system but by simply monitoring and making slight adjustment to the excess air in the burner, good burner performance can be ensured.

Gas 3: Minimum direct efficiency of 61% and maximum indirect efficiency of 84%

1. Bridged when all boiler operating parameters in terms of regular tuning, load management on the boiler, feed water temperature, etc., in the boiler house is done regularly.
2. Maintaining the correct level in the feed water tank for better feed water temperature.

Efficient steam generation

13.1 Introduction

The function of the steam distribution system is to get the steam to where it is needed and return the condensate to the boiler, doing both as efficiently as possible. Distribution heat losses account for 3 to 10% of the total energy generated in a boiler system. Energy management can reduce the heat loss by improving the insulation, detecting and repairing steam and condensate leaks, maintaining the steam traps and condensate pumps and providing water treatment. A well designed steam distribution network can improve the efficiency of the steam systems. For optimum performance of the distribution and steam enduse equipment, a supply of right quantity and quality of steam is of vital importance.

The losses in the steam distribution system can be in the form of:

1. Radiation and convection.
2. Pressure losses in the distribution pipe lines.
3. Steam leaks in joints, valves, gauges, etc.
4. Steam losses due to improper selection, incorrect location, wrong positioning and malfunctioning of traps.
5. Inappropriate location and capacity of air vents.
6. Poor dryness fraction of steam.

Steam losses due to external leakages can easily be identified. Such leakages can be plugged using online sealing techniques. The valves in the bypass around the steam traps as well as malfunctioning steam traps are the prime sources of internal leakages.

These are difficult to detect as they are hidden and invisible in the flash steam. It is, therefore essential to improve the steam distribution system. Following are some important aspects to be taken care of:

1. Properly select, size and maintain the distribution system steam traps.
2. Insulate all distribution system pipes, flanges and valves.
3. Ensure that steam mains are properly laid out, sized, adequately drained and adequately air vented.
4. Ensure that distribution system piping is correctly sized to maintain appropriate system pressure drops.

5. Ensure that distribution system piping is adequately supported, guided and anchored and that appropriate allowances are made for pipe expansion at operating temperatures.

A practical steam distribution system should necessarily compromise between the above ideal conditions and several other factors. Lack of attention to these will significantly increase operating costs, either because of reduction in overall efficiency or increase in maintenance costs or both.

13.2 Energy conservation

13.2.1 Steam piping layout

Steam piping transports steam from the boiler to the end-use services. Important characteristics of a well-designed steam system piping are that it is adequately sized, configured and supported. Installations of larger pipe diameters could be more expensive, but can reduce the pressure drop for a given flow rate and also help to reduce the noise associated with steam flow. Hence, one consideration should be given to the type of environment in which the steam piping will be located when selecting the pipe diameter.

Important configuration issues are flexibility and drainage. Piping, especially at equipment connections, needs to accommodate thermal reactions during systems start-ups and shutdowns. Piping should be equipped with a sufficient number of appropriately sized drip legs to promote effective condensate drainage and should be pitched properly to promote the drainage of condensate to these drip lines. Typically, these drainage points experience two very different operating conditions, viz., normal operation and start-up. Both load conditions should be considered in the initial design.

Mechanical type moisture separators with traps should be provided in piping at interval, to separate the fine moisture particles in the steam. Automatic air vents should be fixed at the dead end of steam mains to allow removal of air/ non-condensable which tends to accumulate in steam space.

13.2.2 Steam pipe sizing and redundancy

Proper sizing of the steam pipelines involves selecting a pipe diameter which gives acceptable pressure drop between the boiler and the user. Pipe sizing can be done either based on the velocity or on the desired pressure-drop. Pipe sizing can be done from the general recommendations on line velocities of different fluid based on the specific volume of steam for the chosen distribution pressure and quality of steam, whether wet or superheated.

The velocities for various types of steams are:

1. Superheated 50–70 m/sec.

2. Saturated 30–40 m/sec.

3. Wet or exhaust 20–30 m/sec.

Unused steam piping experiences the same losses as the rest of the system. It is therefore imperative to isolate the unused steam lines immediately. Pipe routing is made for transmission of steam in the shortest possible way, so as to reduce the pressure drop in the system.

13.2.3   Steam pressure

The steam distribution pressure should be adjusted in accordance with the pressure generated and the pressure required at the consumer side. If steam piping already exists then the pressure should be adjusted for lower operating cost. However, at the designing stage, it is desirable to consider steam distribution at the same pressure at the source, or at a moderately high intermediate pressure; if the generation is at very high pressure. Distribution of the steam at the same pressure that of source has the following advantages:

1. The steam velocity along within the pipes will be lowered and this reduces both noise and erosion.

2. It provides stable pressure at the user end due to lower pressure drop and higher operating margins.

3. The capital cost is reduced as the pipe line is so smaller in size.

Nonetheless, for long distribution systems, it is economical to super-heat the steam to minimise the steam losses. The piping needs to be properly sized and well insulated.

Estimating pressure requirements for small distribution systems is relatively simple, viz., it should just meet the minimum user requirement, unless future expansion of the system or new equipment requiring higher pressures is envisaged. For systems where only a small quantity of high pressure steam is actually required, but where large quantities of low pressure steam are used, the possibility of separating the two should be considered.

13.2.4   Insulation in steam distribution

Heat losses through the surface of the steam distribution pipes can significantly increase energy use and cost. Good engineering design of insulation system will reduce undesirable heat loss and will often improve environmental condition.

Poorly insulated/uninsulated steam distribution and condensate return lines are a constant source of wasted energy. A good and proper insulation can typically reduce energy losses by 90% and help to ensure proper steam quality and pressure at plant equipment. It would also reduce leakages and other issues due to erosions or water hammering due to excessive condensate in steam.

Table 13.1 illustrates steam line losses for non-insulated pipes of different diameters. Table 13.2 gives different types of material used for insulation.

Table 13.1: Steam line losses for non-insulated pipes of different diameters.

Pipe diameter (NB)	Heat loss (kcal/hr for 100 M bare pipe) Steam pressure (kg/cm²g)			
	1.0	10.0	20.0	40.0
25	13210	26892	35384	46706
50	22174	45291	59444	79259
100	39158	80203	105679	141534
200	69824	98130	191543	257121
300	99546	207584	274576	369876

Table 13.2: Different types of material used for insulation.

Material	Density (kg/m³)	Thermal conductivities (W/m°C)				Maximum temperature (°C)
		0°C	100°C	200°C	300°C	
Polystyrol	20–50	0.032				70
Cork	100–200	0.032				80
Glass wool (non fibre)	40–60	0.031	0.050			200
Long fibre	80	0.031	0.048	0.073	0.110	500
Short fibre	100	0.036	0.051	0.051	0.102	700
Rockwool and glass wool	40–250	0.028	0.039			800
Asbestos	80–250	0.042				600

13.2.5 Economical insulation thickness

As the thickness of the insulation increases, the cost of material and installation also goes up. The cost of lost energy, on the other hand, goes down upto a certain thickness. Above this thickness, the gains due to drop in the surface temperature are compensated with increase in the surface area of the insulation. In other words, the energy saving also goes up, but at a slower rate of increase than the cost of the materials and installation.

At a particular point, the total cost, which is the sum of the lost energy and the material cost, reaches a minimum point; which is the economic thickness of insulation. Figure 13.1 illustrates the method of determining the insulation thickness.

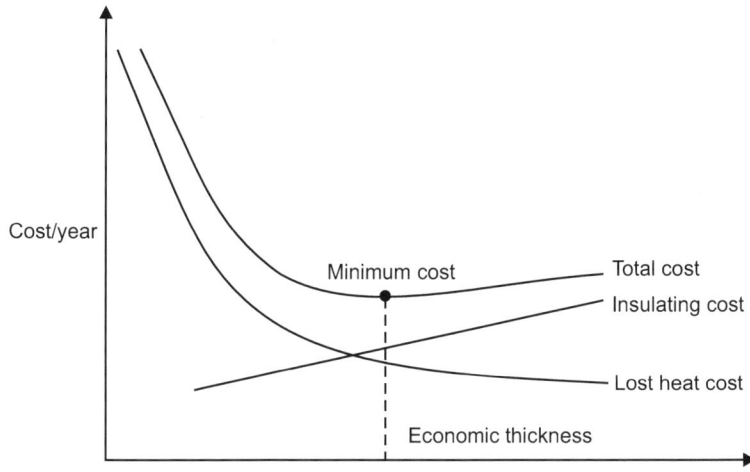

Figure 13.1: Method of determining insulation thickness.

The recommended insulation thickness for mineral wool which is commonly used in various industries is given in Table 13.3.

Table 13.3: Recommended insulation for mineral wool.

Temperature of process fluid (°C)	Diameter of pipe (NB)				Flat surface/ above 200 NB
	Upto 40	50–80	90–125	150–200	
Up to 90	25	25	25	40	40
91–150	40	40	50	50	65
151–250	65	65	75	75	90
251–350	75	75	100	100	100
351–450	90	90	100	115	125
451–550	90	100	115	125	140
551–650	90	100	115	130	150

Major factors determining insulation selection are:
1. Operating temperature.
2. Thermal conductivity of the insulating material.
3. Resistance to heat, weather and adverse atmospheric conditions.
4. Ability to withstand vibration, noise and mechanical damage.
5. Resistance to chemicals/environment.
6. Resistance to fire.
7. Extent of shrinking or creaking during use.
8. Jacking for insulation.

13.2.6 Steam traps and strainers

As steam moves throughout the system, it looses a small part of heat through surfaces, due to condensation. The condensate, travelling at over 200 km/hr may erode the pipe lines (especially at bends/partially open valves) and even lead to water hammering and can damage equipment; if not removed effectively. Steam traps are automatic valves that separate condensate from the steam. A leaky trap wastes energy by allowing steam to enter the condensate return. A malfunctioning trap may not expel the condensate from the steam line, thus reducing efficiency of the system. Steam traps are classified into three main groups–mechanical, thermostatic and thermodynamic, with several different types of traps in each group. It is essential to understand the operations and functions of the traps and carefully read the instructions, since the traps operate in different ways and sizes, as positioning and installation procedures vary.

13.2.7 Selection of steam traps

Selection of an appropriate trap should be made based on the capacity curve of the trap, which varies with every manufacture. Due care must be taken to understand loading, normal as well start up and differential pressure across the trap, type of applications, possibility of non-condensable in condensate, of the capacity curve.

13.2.8 Steam trap leakage

Leakage in the steam traps allows the steam to blow into the condensate system which is then vented to the atmosphere. A regular inspection must be carried out for steam traps and valves and leaks should be attended to immediately. To emphasise on controlling leakages from the steams traps and orifices, the losses at different sizes and their pressure are shown in Table 13.4.

Table 13.4: Losses at different sizes and their pressure.

Steam pressure kg/cm²	Orifice size inch	Steam losses kg/hr
3.5	1/8	18
	1/4	72
	1/2	290
5.5	1//8	26
	1/4	100
	1/2	400
15	1/8	60
	1/4	240
	1/2	960

(Cont'd...)

Steam pressure kg/cm²	Orifice size inch	Steam losses kg/hr
42	1/8	150
	1/4	600
	1/2	2400

13.2.9   Strainers

Performance of steam traps decreases due to dirt and scale accumulation. To eliminate this problem it is essential to install pipe-line strainer. Strainers are fitted before the traps, if the traps do not have built-in trap strainer. It is advisable to check the strainers at regular intervals.

13.2.10 Water hammer

One of the most common complaints is that a system sometimes develops a hammer-like noise commonly referred to as water hammer. Water hammer in steam lines is normally caused by the accumulation of condensate. It may indicate a condition which could produce serious consequences including damaged vents, traps, regulators and piping.

Two types of water hammer can occur in steam systems:

1. The first type is usually caused by the accumulation of condensate (water) trapped in a portion of horizontal steam piping. The velocity of the steam flowing over the condensate causes ripples in the water. Turbulence builds up until the water forms a solid mass, or slug, filling the pipe. This slug of condensate can travel at the speed of the steam and will strike the first elbow in its path with a force comparable to a hammer blow. In fact, the force can be great enough to break the back of the elbow.

2. The second type of water hammer is actually cavitation. This is caused by a steam bubble forming or being pushed into a pipe completely filled with water. As the trapped steam bubble looses its latent heat, the bubble collapses, the wall of water comes back together and the force created can be severe. This condition can crush float balls and destroy thermostatic elements in steam traps. Cavitation is the type of water hammer that usually occurs in condensate return lines or pump discharge piping.

Precautions to prevent water hammer in steam lines are:

1. Steam pipes must be pitched away from the boiler towards a drip trap station. Drip trap stations must be installed ahead of any risers, at the end of the main and every 100 to 150 m along the steam piping.

2. Drip traps must be installed ahead of all steam regulator valves to prevent the accumulation of condensate when the valve is in a closed position.

3. 'Y' strainers installed in steam lines should have screen and dirt pocket mounted horizontally to prevent condensate from being collected in the screen area and being carried along in slugs when steam flow occurs.

4. All equipment using a modulation regulator on the steam supply must provide gravity condensate drainage from the steam traps. Lifts in the return line must be avoided.

Another type of loss observed in steam traps, occurs when hot or pressurised condensate passes through the traps and the water flashes off a certain percentage of the steam due to instant pressure change.

This is called 'Flash steam loss'. In this case it is recommended to use pressurised condensate recovery system and/or flash steam recovery system for complete recovery of thermal energy.

13.2.11 Steam quality

The quality of steam also determines the performance of steam distribution system. Good quality steam means dry moisture-free steam, free from air, carbon dioxide and other non-condensable matter.

13.2.12 Moisture in steam

Saturated steam generated in packaged boiler contains 2 to 5% moisture; while steam from coil type non-IBR boilers could have 10 to 60% moisture. Super-heated steam on the other hand does not contain any moisture. However, some moisture is picked up while de-superheating the steam. The steam also gives away latent heat and becomes wet, while being transported through distribution system. The wet steam contains particles of water droplet which have not evaporated. These droplets do not contribute to heat transfer and it is essential to remove it from the steam.

A moisture separator at the entrance of the equipment separates the droplets and drains them through traps. The wetness can also be reduced by resorting to pressure reduction of steam prior to its use.

13.2.13 Non-condensable in steam

Dissolve oxygen in the boiler feed water, if not removed properly, gets carried away with the steam. The bicarbonate salts in the feed water generate carbon dioxide which is also transported with the steam. The problems associated with non-condensable can be summed up as under.

1. Reduction in the heat transfer area to the extent of space occupied.

2. Drop in heat transfer rate due to reduction in effective steam temperature (based on the partial pressure of the steam in the steam air mixture).

3. Additional resistance to heat transfer due to formation of barrier layer.

It is therefore very important to remove the non-condensables through air vents provided at proper location and also installing appropriate type of steam trap.

13.2.14 Salient points in steam generation

1. Replace damaged/wet insulation.
2. Avoid steam leakages.
3. Provide dry steam for process.
4. Utilising steam at lower acceptable pressure for the process.
5. Ensure proper utilisation of directly inject steam.
6. Minimise heat transfer.
7. Use condensate recovery system.
8. Insulate all steam pipelines and hot process equipment.
9. Recover flash steam.
10. Maintain at least 125 mm per meter of falling slope for steam piping.
11. Provide drain points at lower points in the main and where the steam main rises.
12. Drain points in the main lines should be through an equal tee connection only.
13. The branch lines from the mains should always be connected at the top.
14. Insure supports as well as an alternation in level can lead to formation of water pockets in steam, leading to wet steam delivery.

Thus, steam is common and convenient mode to convey energy and is used in almost all major industrial processes. Steam economy greatly depends on delivering the steams through properly designed steam distribution lines. It is essential to adopt all possible measures including new technologies to optimise the steam distribution costs.

13.3   Improving efficiency of steam systems

Steam is the most popular heat transfer medium for process industry. However, the inefficiencies inbuilt in the design and operation of steam systems, offer great scope for energy saving. Fundamentally it involves reducing losses and recovering as much heat as possible.

1. Steam is generated by evaporation of water.
2. The process involves high heat absorption-hence plant sizes and costs are not impracticably large. Water itself is by far the most common liquid on earth and therefore plentiful and cheap. It is also chemically stable and non-hazardous to health.

3. Steam supplies heat to any process by condensing at a constant temperature and with high heat transfer coefficients. This constant condensing temperature eliminates any temperature gradient over the heat transfer surfaces.

4. Steam does not need any circulating pump to carry it to the usage points. Steam pressure has a direct relation to its temperature.

All of the above ensure that steam is the most popular heat transfer medium for process industry. However, the flexibility offered by the steam system often results in large inefficiencies inbuilt in the design and operation. That is why most process industries offer great scope for fuel savings in steam systems.

In order to identify this scope lets take a look at the energy balance of a very simple steam system, having one boiler, a short distribution network and a single indirectly heated consuming point Fig. 13.2.

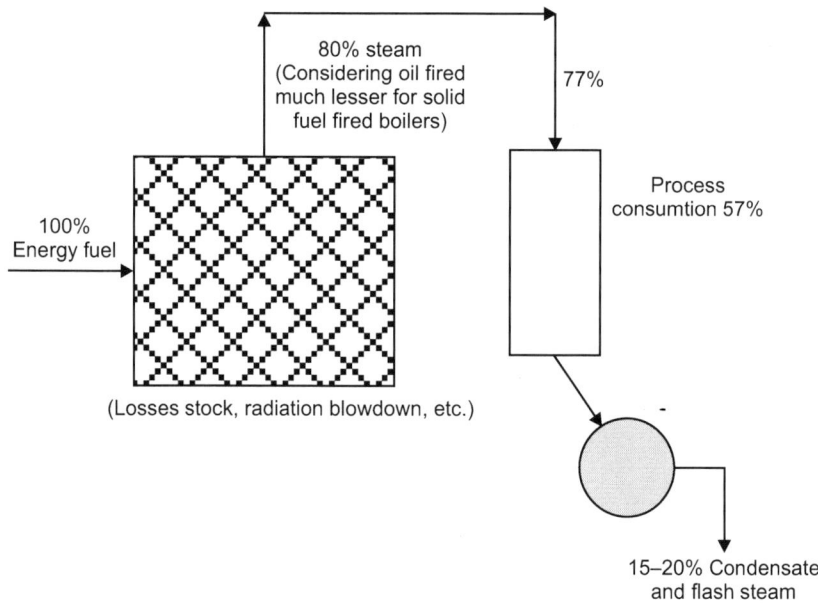

Figure 13.2: Energy balance of simple steam system.

As seen in the diagram, only about 60% of the fuel energy burnt in the boiler is actually useful for the heat transfer in process.

This is what makes a good steam and condensate system design. Following practices are mandatory for optimising fuel consumption.

By designing a proper system, one can ensure:

1. Reducing the possibility of actual losses-by achieving better combustion efficiency in boiler, insulating distribution network, plugging leaks, keeping by-passes closed, etc.

2. Recovering as much heat as possible - from boiler stack, condensate and flash steam, etc.

Proper specification for quality of steam for process heating determines the actual demand and hence affects the mass balance directly.

This section focuses only on steam generation which is the heart of the steam and condensate loop.

13.3.1 Optimising steam generation

This is a case study for furnace oil fired boilers based on a survey involving a sample size of 30 process plants for furnace oil fired boilers:

Efficiency %	Best	Average	Worst
Direct	82	70	60
Indirect	84	80	75

13.3.2 Understanding boiler efficiency

$$\text{Direct efficiency} = \frac{\text{Output steam energy}}{\text{Input fuel energy}}$$

$$= \frac{\text{Steam generation (kg/hr)} \times (\text{Hs} - \text{Hw})}{\text{Fuel consumption (kg/hr)} \times \text{GCV of fuel}}$$

13.3.3 Boiler efficiency

Thus, direct efficiency translates into S:F ratio and determines the fuel consumption for a given loading (Fig. 13.3) on the boiler, steam pressure and feed water temperature.

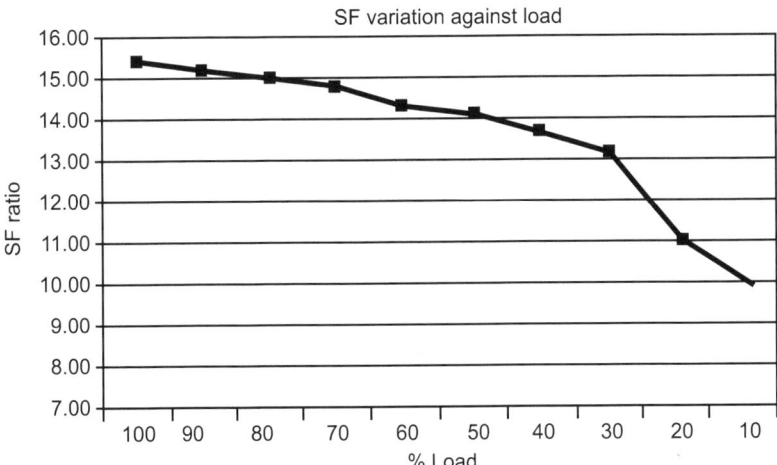

Figure 13.3: SF ratio v/s load.

Direct efficiency is averaged normally over a period of time, batch, day, month, etc.

Indirect efficiency - as per BS-845, boiler efficiency is calculated by deducting losses: = 100 − (sum of all losses in %)
where, some of the losses are:

1. Loss due to sensible heat in dry flue gas (stack loss).
2. Loss due to enthalpy in the water vapour in the flue gases (enthalpy loss).
3. Radiator loss.
4. Unburnt fuel losses.

Thus, indirect efficiency directly translates into corrective actions required to be taken. Normally, indirect efficiency is measured during a spot check and not averaged over a period of time and hence may give misleading results. Difference between direct and indirect efficiency is given Table 13.5.

Table 13.5: Difference between direct and indirect efficiency.

Direct efficiency	*Indirect efficiency*
Being a ratio, it is normally calculated over a period of time.	Can be calculated after capturing combustion parameters during a spot check.
Practically computed by measuring actual flow of fuel to the boiler and steam generated from the boiler for a period of time.	Computed by measuring combustion parameters like stack temperature, excess air (O_2 % in stack), CO_2, CO, etc., during a spot check.
Takes into account issues related to house-keeping, like oil spillage, handling losses, start-stop losses which spot checks of indirect efficiency.	Is an indication of combustion properties and hence helps in better tuning of the burner, dampers, etc., to achieve better performance.

Some of the reasons why direct efficiency was found to be lower than indirect in the sample survey are:

1. As mentioned above, direct efficiency establishes the performance over a period of time as against spot checks of indirect efficiency. It takes into account actual measurements of input and output to the boiler for a specified time period. Thus, partially the difference is accounted for, as indirect efficiency parameters are not maintained uniformly over a period of time.
2. The biggest contributor for loss of direct efficiency was observed to be:
   (a) The boiler loading and its turndown. Many of the boilers were observed to be grossly oversized and showed poor results when operated at partial or low loads.
3. Another reason for poor direct efficiency was attributed to start/stop losses, i.e., the frequent tripping of burner due to pressure being achieved.

4. Lastly, losses due to blow down, oil leakages from burner assembly, steam leakage or venting (safety valve) from the boiler, etc., are all considered while computing direct efficiency-but not while computing indirect efficiency.

13.3.4   Corrective actions

1. The first step is proper diagnostics - online monitoring of the boiler efficiency - direct as well as indirect.
2. Proper selection and tuning of the burner.
3. Derating of the boiler, if oversized (only radiation losses cannot be prevented and will continue to be higher).
4. Proper pressure switch setting to prevent frequent on/off and safety valve venting or manually setting the burner on low fire instead of auto modulation.
5. Better housekeeping to avoid leakages and losses.

Benefits and savings

If direct efficiency improves from 70% on an average to 82% which is the best, fuel consumption will reduce by 17% - other parameters like feed, water, temperature, etc., remaining same.

13.3.5   Waste heat recovery in textile industry

Once all of the above factors are taken care of and operating parameters well established and under control, one should worry about:

1. Complete condensate and flash steam recovery. In a typical system this itself can reduce fuel consumption further by 10–15%.
2. Waste heat recovery from exhaust flue gases by way of economisers (water pre-heaters) or air pre-heaters which can further reduce fuel consumption by 3–5%.

Apart from generation and condensate/flash steam recovery, equal opportunity exists in optimising steam consumption in process equipment and distribution network, which is more dependent on the type of industry and process. Unique packages have been designed to provide online energy/mass balance on steam and condensate loop, for various industries.

13.4   Energy-efficiency improvement opportunities in steam systems of textile plants

Steam systems are often found in textile plants and can account for a significant amount of end-use energy consumption. Improving boiler efficiency and capturing excess heat can result in significant energy savings and improved production. Common performance improvement opportunities for the generation and distribution of industrial steam systems are given below.

13.4.1 Steam generation in textile industry

Demand matching

A boiler is more efficient in the high-fire setting. Since process heating demands may change over time, situations can occur in which a boiler is operating beneath its optimum efficiency. Also, boilers may have been oversized, for instance because of additions or expansions that never occurred. Installing energy conservation or heat recovery measures may also have reduced the heat demand. As a result, a facility may have multiple boilers, each rated at several times the maximum expected load. An additional common problem with oversized boiler is the boiler 'short cycling', which occurs when an oversized boiler quickly satisfies process or space heating demands and then shuts down until heat is again required. Fuel savings can be achieved by adding a smaller boiler to a system, sized to meet average loads at a facility. Multiple small boilers offer reliability and flexibility to operators to follow load swings without over-firing and short cycling. In particular, facilities with large seasonal variations in steam demand should use to operate small boilers when demand drops, rather than operating their large boilers year-round. Operation measures to operate boilers on the high-fire setting have a average payback time of 0.8 years and the installation smaller boilers to increase the high-fire duty cycle has a average payback time of 1.9 years.

Boiler allocation control

Systems containing multiple boilers offer energy-saving opportunities by using proper boiler allocation strategies. This is especially true if multiple boilers are operated simultaneously at low-fire conditions.

Automatic controllers determine the incremental costs (change in steam cost/change in load) for each boiler in the facility and then shift loads accordingly. This maximises efficiency and reduces energy costs. If possible, schedule loads to help optimise boiler system performance.

The efficiency of hot water boilers can improve through use of automatic flow valves. Automatic flow valves shut off boilers that are not being used, preventing the hot water from the fired boiler getting cooled as it passes through the unused boilers in the system. Where valves are left open the average flow temperature is lower than designed for and more fuel is used.

Flue shut-off dampers

Where boilers are regularly shut down due to load changes, the heat lost to the chimney can be significant. A solution to stop this loss of hot air is to fit fully closing stack dampers, which only operate when the boiler is not required. Another alternative is to fit similar gas tight dampers to the fan intake.

Maintenance

In the absence of a good maintenance system, the burners and condensate return systems can wear or get out of adjustment. These factors can end up costing a steam system up to 20–30% of initial efficiency over 2–3 years. A simple maintenance programme to ensure that all components of the boiler are operating at peak performance can result in substantial savings and furthermore reduce the emission of air pollutants.

On average the possible energy savings are estimated at 10% of boiler energy use. The establishment of a maintenance schedule for boilers has an average payback time of 0.3 years.

Insulation improvement

The shell losses of a well-maintained boiler should be less than 1%. New insulation materials insulate better and have a lower heat capacity. As a result of this lower heat capacity, the output temperature is more vulnerable to temperature fluctuations in the heating elements. Improved control is therefore required to maintain the output temperature range of the old fire brick system. Savings of 6–26% can be achieved by combining improved insulation with improved heater circuit controls.

Reduce fouling

Fouling of the fireside of the boiler tubes and scaling waterside of the boiler should be controlled. Tests show that a soot layer of 0.03 inches (0.8 mm) reduces heat transfer by 9.5%, while a 0.18 inch (4.5 mm) layer reduces heat transfer by 69%. Scale deposits occur when calcium, magnesium and silica, commonly found in most water supplies, react to form a continuous layer of material on the waterside of the boiler heat exchange tubes.

Tests showed that for water-tube boilers 0.04 inches (1 mm) of buildup can increase fuel consumption by 2%.

In fire-tube boilers scaling can lead to a fuel waste up to 5%. Moreover, scaling may result in tube failures.

Scale removal can be achieved by mechanical means or acid cleaning. The presence of scale can be indicated by the flue gas temperature or be determined by visually inspecting of the boiler tubes when the unit is shut down for maintenance. Fouling and scaling are more of a problem with coal-fed boilers than natural gas or oil-fed ones (i.e., boilers that burn solid fuels like coal should be checked more often as they have a higher fouling tendency than liquid fuel boilers do).

Optimisation of boiler blow down rate

Insufficient blow down may lead to carryover of boiler water into the steam, or the formation of deposits. Excessive blow down will waste energy, water and chemicals. The optimum blow down rate is determined by various factors including the boiler type, operating pressure, water treatment and quality of makeup water. Blow down rates typically range from 4% to 8% depending on boiler feed water flow rate, but can be as high as 10% when makeup water has a high solids content. Minimising blow down rate can therefore substantially reduce energy losses, makeup water and chemical treatment costs. Optimum blow down rates can be achieved with an automatic blow down control system. In many cases, the savings due to such a system can provide a simple payback of 1 to 3 year.

Reduction of flue gas quantities

Often, excessive flue gas results from leaks in the boiler and the flue, reducing the heat transferred to the steam and increasing pumping requirements. These leaks are often easily repaired. This measure consists of a periodic repair based on visual inspection or on flue gas monitoring.

Reduction of excess air

The more air is used to burn the fuel, the more heat is wasted in heating air. Air slightly in excess of the ideal stoichometric fuel/air ratio is required for safety and to reduce NO_x emissions and is dependent on the type of fuel. Poorly maintained boilers can have up to 140% excess air leading to excessive amounts of waste gas. An efficient natural gas burner however requires 2% to 3% excess oxygen, or 10% to 15% excess air in the flue gas, to burn fuel without forming carbon monoxide. A rule of thumb is that boiler efficiency can be increased by 1% for each 15% reduction in excess air. Fuel-air ratios of the burners should be checked regularly. On average the analysis of proper air/fuel mixture had a payback time of 0.6 years.

Flue gas monitoring

The oxygen content of the exhaust gas is a combination of excess air (which is deliberately introduced to improve safety or reduce emissions) and air infiltration (air leaking into the boiler). By combining an oxygen monitor with an intake airflow monitor, it is possible to detect (small) leaks. Using a combination of CO and oxygen readings, it is possible to optimise the fuel/air mixture for high flame temperature (and thus the best energy-efficiency) and low emissions. The payback of installing flue gas analysers to determine proper air/fuel ratios on an average is 0.6 years.

Pre-heating boiler feed water with heat from flue gas (economiser)

Heat from flue gases can be used to pre-heat boiler feed water in an economiser. By pre-heating the water supply, the temperature of the water supply at the inlet to the boiler is increased, reducing the amount of heat necessary to generate steam thus saving fuel. While this measure is fairly common in large boilers, there often still is potential for more heat recovery. Generally, boiler efficiency can be increased by 1% for every 22°C reduction in flue gas temperature. By recovering waste heat, an economiser can often reduce fuel requirements by 5% to 10% and pay for itself in less than 2 years.

Recovery of heat from boiler blow down

When the water is blown from the high-pressure boiler tank, the pressure reduction often produces substantial amounts of steam. Up to 80% of the heat in the discharge is recoverable by using flash vessels and heat exchangers. The recovered heat can subsequently be used for space heating and feed water pre-heating increasing the efficiency of the system. Any boiler with continuous blow down exceeding 5% of the steam rate is a good candidate for the introduction of blow down waste heat recovery. If there is a non-continuous blow down system, then consider the option of converting it to a continuous blow down system coupled with heat recovery. Larger energy savings occur with high-pressure boilers. The use of heat from boiler blow down on average has payback period of 1.6 years.

Recovery of condensate

By installing a condensing economiser, companies can improve overall heat recovery and steam system efficiency by up to 10%. Many boiler applications can benefit from this additional heat recovery. Condensing economisers require site-specific engineering and design and a thorough understanding of the effect they will have on the existing steam system and water chemistry.

Hot condensate can be returned to the boiler saving energy and reducing the need for treated boiler feed water as condensate, being condensed steam, is extremely pure and has a high heat content. Increasing the amount of returned condensate has an average payback period of 1.1 years. Condensate has also been used to provide for hot water supply. This measure had an average payback period of 0.8 years.

Combined heat and power (CHP)

Combined heat and power (CHP) or cogeneration is the sequential production of two forms of useful energy from a single fuel source. In most CHP applications, chemical energy in fuel is converted to both mechanical and thermal energy. The mechanical energy is generally used to generate electricity, while the thermal energy or heat is used to produce steam, hot water, or hot air.

CHP systems have the ability to extract more useful energy from fuel compared to traditional energy systems such as conventional power plants that only generate electricity and industrial boiler systems that only produce steam or hot water for process applications.

CHP provides the opportunity to use internally generated fuels for power production, allowing greater independence of grid operation and even export to the grid. This increases reliability of supply as well as the cost-effectiveness. In addition, transportation losses are minimised when CHP systems are located at or near the end user.

The cost benefits of power export to the grid will depend on the regulation in the country where the industry is located, but can provide a major economic incentive. Not all countries allow wheeling of power (i.e., sales of power directly to another customer using the grid for transport) and for the countries that do allow wheeling, regulations may also differ with respect to the tariff structure for power sales to the grid operator.

13.4.2   Steam distribution system

Shutting off excess distribution lines

Installations and steam demands change over time, which may lead to under-utilisation of steam distribution capacity utilisation and extra heat losses. It may be too expensive to optimise the system for changed steam demands. Still, checking for excess distribution lines and shutting off those lines is a cost-effective way to reduce steam distribution losses.

Proper pipe sizing

When designing new steam distribution systems it is very important to account for the velocity and pressure drop. This reduces the risk of oversizing a steam pipe, which is not only a cost issue but would also lead to higher heat losses. A pipe that is too small may lead to erosion and increased pressure drop.

Insulation related measures

Insulation can typically reduce energy losses by 90% and help ensure proper steam pressure at plant equipment. The application of insulation can lead to significant energy cost savings with relatively short payback periods. For instance, the average payback period of the insulation on steam and hot water lines is 1.0 years, that of condensate lines 1.1 year and that of the feedwater tank 1.1 years.

Checking and monitoring steam traps

A simple programme of checking steam traps to ensure they operate properly can save significant amounts of energy. If the steam traps are not maintained for 3 to 5 years, 15–30% of the traps can be malfunctioning, thus allowing live steam to escape into the condensate return system. In systems with a regularly scheduled maintenance programme, leaking traps should account for less than 5% of the trap population.

The repair and replacement of steam traps has an average payback time of 0.4 years. Energy savings for a regular system of steam trap checks and follow-up maintenance is estimated to be up to 10%.

Thermostatic steam traps

Using modern thermostatic element steam traps can reduce energy use while improving reliability. The main advantages offered by these traps are that they open when the temperature is close to that of the saturated steam (within 2°C), purge non-condensable gases after each opening and are open on startup to allow a fast steam system warm-up. These traps are also very reliable and useable for a large range of steam pressures. Energy savings will vary depending on the steam traps installed and the state of maintenance.

Shutting of steam traps

Other energy savings can come from shutting of steam traps on superheated steam lines when they are not in use. This has an average payback time 0.2 years.

Reduction of distribution pipe leaks

As with steam traps, the distribution pipes themselves often have leaks that go unnoticed without a programme of regular inspection and maintenance. On average leak repair has a payback period of 0.4 years.

Recovery of flash steam

When a steam trap purges condensate from a pressurised steam distribution system to ambient pressure, flash steam is produced. Depending on the pressures involved, the flash steam contains approximately 10% to 40% of the energy content of the original condensate. This energy can be recovered by a heat exchanger and used for space heating or feed water pre-heating.

The potential for this measure is site dependent, as it is unlikely that a plant will build an entirely new system of pipes to transport this low-grade steam, unless it can be used close to the steam traps. Sites using multi-pressure steam systems can route the flash steam formed from high-pressure condensate to reduced pressure systems.

Prescreen coal

In some textile plants in which a coal-fired boiler is used (which is common in China and some other countries) raw coal is fed into the boiler and burned on stoke chains, which allows small-sized coal to pass through the chain and become wasted. To address this loss, companies should adopt spiral coal screen technology to screen the raw coal. This device greatly increases the rate of separation of good and bad quality coal, preventing small coal particles from falling off the stoke chain and increasing the calorific value of the fired coal. The installation of a spiral coal screener could save about 79 kg coal/T fabric (1.8 GJ/T fabric) in wet-processing plants based on the case-studies in China. It has an estimated cost of $35,000. The screener would pay back its cost in about five months.

Waste heat recovery in textile industry

14.1   Introduction

Heat recovery systems are designed to conserve energy by reusing available waste heat. They transfer heat from sources of waste heat to uses for the recovered heat, with various types of heat recovery equipment. The conservation of energy not only reduces our dependence on imported fuels, but produces a cost saving to pay back the system cost. The cost saving increases as fuel costs increase, creating an inflation-proof investment.

14.2   Basic heat recovery in textile industry

The elements of a heat recovery system is shown in Fig. 14.1. The 'source' produces waste heat as a result of a process or building operation. The waste heat can be contained in a gas, liquid or vapour. Its temperature may be very high, as in the exhaust from a furnace or it may be close to ambient temperature, as in the exhaust from a building ventilator.

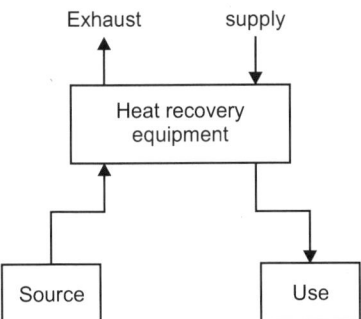

Figure 14.1: Elements of a heat recovery system.

The use consumes heat as part of its operation. Besides liquid, gas and vapour, the use can encompass process materials pre-heated before entering the process. The heat recovery equipment indicates a means for transferring the waste heat from the source in a form acceptable by the use. The type of device used for the heat recovery equipment depends upon the nature of the source and use and their respective temperatures. Examples of heat recovery equipment include heat exchangers, waste heat boilers and boiler economisers. Following the heat recovery equipment, the exhaust fluid is either vented or drained.

The supply is received from outside sources, return from the process or building return air. A simple example of a heat recovery system is shown in Fig. 14.2. As shown, exhaust gases from the boiler stack pass through an economiser in the stack. The economiser reduces the gas temperature. Returning boiler feedwater is heated in the economiser before entering the boiler. The cost saving produced by the stack economiser appears as reduced fuel to pre-heat the boiler feedwater.

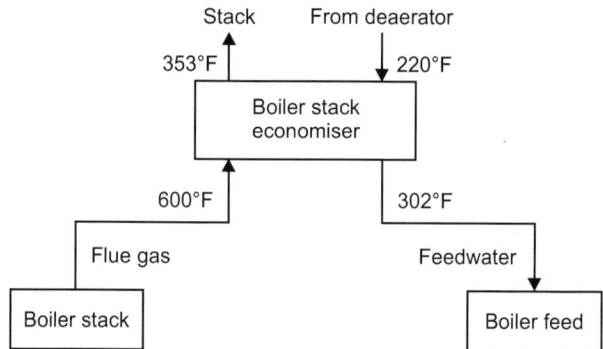

Figure 14.2: Boiler feedwater heat recovery system.

The example of Fig. 14.2 illustrates the principles of heat recovery:

1. The exhaust temperature falls as a result of losing recoverable waste heat.
2. The supply temperature rises as a result of using recovered waste heat.
3. The amount of recovered waste heat given up by the exhaust must equal the amount of recovered waste heat gained by the supply.
4. The amount of recovered waste heat is less than the total amount of exhaust heat.

Since the presence of heat in a material is measured by the material's temperature, the material's loss or gain of heat is reflected by its change in temperature. The greater the amount of heat loss or gain, the greater the change in temperature. The heat loss from the exhaust must appear in the heat gained by the supply, based upon the basic laws of thermodynamics. There is no other place for the heat to exist. The temperatures of the exhaust and supply are not necessarily the same because they may originate under very different conditions, but the amount of heat lost and gained is always the same. Finally, the heat recovery equipment can recover only a portion of the total heat in the exhaust. It can reduce the exhaust temperature to the extent permitted by the supply and exhaust temperatures and by its design. The supply temperature is the lowest theoretical exhaust temperatures leaving the equipment, because the supply cannot cool the exhaust below its own temperature. The equipment design fixes how closely the actual exhaust temperature approaches the supply temperature. Any remaining source heat after the heat recovery equipment is lost in the exhaust leaving the heat recovery equipment

14.2.1   Benefits of heat recovery

Heat recovery benefits fall into three categories:

1. Reduction of energy cost.
2. Reduction of equipment cost and size.
3. Reduction of energy use.

Reduction of energy cost is the primary benefit of heat recovery. Any heat recovered from the exhaust and returned to the supply need not be supplied by purchase energy. Further, any increases in energy prices result in increase heat recovery benefits. A heat recovery system is both inflation and price increase-proof. Very few other investments are so free of economic risk.

Additional cost benefits for heat recovery systems are available as equipment cost and size reduction. The use of recovered heat reduces the amount of heat furnished from purchased energy. Oil and gas supply pipes, electrical facilities, burners, boilers and support structures often can be reduced. If standby facilities are required, temporary, rather than permanent, equipment can be provided, preserving the cost reduction. The reduction of heat requirements can permit greater utilisation of existing process or ventilation equipment. Increased amounts of product or ventilation can be handled without increasing energy use and without new equipment. Where cyclical or peaking conditions are present, heat recovery allows a flexible way of accommodating periods of high heat demand without providing additional heating facilities.

14.3   Process heating in textile industry

Process heating is a significant source of energy consumption in the industrial and manufacturing sectors and it often results in a large amount of waste heat that is discharged into the atmosphere. Industrial waste heat refers to energy that is generated in industrial processes without being put to practical use. Waste heat losses arise both from equipment inefficiencies and from thermodynamic limitations on equipment and processes.

Industrial process heat recovery effectively recycles this waste heat, which typically contains a substantial amount of thermal energy. The benefits of heat recovery include improving system efficiency, reducing fuel consumption and reducing facility air emissions. While the type and cost-effectiveness of a heat recovery system are dependent on the process temperature and the facility's thermal requirements, many heat recovery techniques are available across low, medium and high temperature ranges.

Process heating refers to the application of thermal energy to a product, raising it to a certain temperature to prepare it for additional processing, to change its properties, or for some other purpose. The energy required for process heating accounts for approximately 20% of total industrial energy use in the U.S., Europe and other development countries. As energy costs continue to rise, facilities are constantly in need of ways to improve the performance of their process heating systems and to reduce their energy consumption.

In many fuel-fired heating systems the exhaust gas that is emitted through a flue or stack is the single greatest heat loss. Process heat recovery saves energy by reusing this otherwise lost heat for a variety of thermal loads, such as pre-heated combustion air, boiler feedwater and process loads, as well as for steam generation.

14.4 Industrial process heat recovery in textile industry

Process heat recovery involves intercepting the waste streams before they leave the plant, extracting some of the heat they contain and recycling that heat.

14.4.1 Applications

Heat recovery can be applied in a wide range of industries. For example, the pulp and paper industry can utilise heat recovery through several processes, from pre-heating milling water with steam to cooling effluent wastewater before sending it to waste treatment. The chemical industry can apply heat recovery to most processes, including chemical manufacturing, as well as to emissions control devices such as recuperative and regenerative thermal oxidisers. The petroleum industry can use heat recovery from production water and glycol regenerators, as well as using heat exchangers between wet and dry crude, in the natural gas cleaning process and in waste treatment operations. The food and beverage industry can achieve savings through the installation of heat exchangers for food pasteurisers, blanch water heat recovery, boiler blow down heat recovery, heating feedstock in the distillation process and recovery of waste heat from dryers and cookers. In both the commercial and industrial sectors, Combined Heat and Power (CHP) systems can efficiently generate electrical power on-site as well as recovering waste heat to generate hot water or steam for process operations.

14.4.2 Benefits of waste heat recovery

The benefits of heat recovery are multiple: economic, resource (fuel) saving and environmental.

First, recovered heat can directly substitute for purchased energy, thereby reducing the facility's energy consumption and its associated costs; further, waste heat substitution can lower capacity requirements for energy generating equipment, thus reducing capital costs for new installation projects.

Second, for a specific heating process, fuel efficiency can be improved through the use of heat recovery, thus reducing the cost of operation. For example, the use of exhaust gas from a fuelfired burner to pre-heat the combustion air can reduce heating energy use by as much as 30%.

Third, due to improved equipment efficiency, smaller equipment capacity requirements and reduced fuel consumption, heat recovery can produce environmental benefits through reductions in emissions of greenhouse gases and atmospheric pollutants.

Sources and quality of waste heat

Waste heat sources can be classified by temperature range, as shown in Table 14.1.

Table 14.1: Waste heat temperature categories.

Category	Temperature (°F)
High	1100 to 3000
Medium	400 to 1100
Low	80 to 400

About 92% of process heat energy used by industry is directly provided by fossil fuels. The waste heat generated from direct-fired processes falls in the high and medium temperature ranges. In the high temperature range, sources of waste heat include refining furnaces, steel heating furnaces, glass melting furnaces and solid waste incinerators.

In the low-temperature range, sources of waste heat include process steam condensate, cooling water from refrigeration condensers, welding machines, boilers and air compressors. In some applications low-temperature waste heat can be used for pre-heating through heat exchangers. For example, cooling water from a battery of spot welders can be used to pre-heat the ventilating air for winter space heating. In the medium temperature range, sources of waste heat include exhaust gases from steam boilers, gas turbines, reciprocating engines, water heating boiler furnaces, fuel cells and drying and baking ovens. Potential heat recovery opportunities include, among others, low pressure steam generation and incoming product pre-heating.

High-temperature waste heat is the highest quality and most useful because it provides more heat recovery options and thus greater potential cost-effectiveness than lower temperature waste heat. It can be made available to do work through the utilisation of steam turbines or gas turbines to generate energy in a cogeneration plant. Table 14.2 lists examples of waste heat sources and their potential applications.

Table 14.2: Various waste heat sources and applications.

Sources	Temperature range	Application
Exhaust gas from refining furnaces, steel heating furnaces, glass melting furnaces, solid waste incinerators	High	Hazardous gas reduction Steam generation Water heating Water pre-heating Combustion air pre-heating Power generation
Exhaust from gas turbines, reciprocating engines, incinerators, furnaces Steam boiler blown down	Medium	Pre-heating incoming product Steam generation Water heating Water pre-heating Combustion air pre-heating
Exhaust gas from fuel burner Reciprocating engine jacket cooling Waste stream from condensers, boilers and air compressors	Low	Absorption cooling Dehumidification Feedwater pre-heating Space heating Evaporation

14.5   Energy conservation in textile industry

A textile or cloth is a flexible material consisting of a network of natural or artificial fibres (yarn or thread). Yarn is produced by spinning raw fibres of wool, flax, cotton, or other material to produce long strands. Textiles are formed by weaving, knitting, crocheting, knotting, or felting. The textile industry or apparel industry is primarily concerned with the design and production of yarn, cloth, clothing and their distribution. The raw material may be natural, or synthetic using products of the chemical industry.

14.5.1   Energy consumption in textile industry

Electricity is the major type of energy used in spinning plants, especially in cotton spinning systems. If the spinning plant just produces raw yarn in a cotton spinning system and does not dye or fix the produced yarn, the fuel may just be used to provide steam for the humidification system in the cold seasons for pre-heating the fibres before spinning them together. Therefore, the fuel used by a cotton spinning plant highly depends on the geographical location and climate in the area where the plant is located.

Wet-processing is the major energy consumer in the textile industry because it uses a high amount of thermal energy in the forms of both steam and heat. The energy used in wet-processing depends on various factors such as the form of the product being processed (fibre, yarn, fabric, cloth), the machine type, the specific process type, the state of the final product, etc.

The wet processing operation consumes almost 50% of the total energy requirement of a composite textile mill. This is an attributable that wet processing operation involves heating of large quantities of water, drying of wet fabric, high temperature such as heat setting, high temperature, dyeing and curing operation, etc.

Steam plays a vital role in the wet processing of fabrics not only due to the fact that steam cost is more than 30% of the total processing cost. In spite of such an important role of steam for cloth processing, generally the aspects of steam generation, distribution and the utilisation are operating at a lower efficiency due to vintage systems in most of the textile mills.

14.6 Waste heat recovery in textile industry

By implementing the waste heat recovery methods we can conserve the energy in the textile industries. The improvements in the boiler blow down, condensate recovery, feed water management and waste water recovery will minimise the energy losses and improve the performance of the thermal systems in textile industries. As the industrial sector continues efforts to improve its energy efficiency, recovering waste heat losses provides an attractive opportunity for an emission free and less costly energy resource.

Textiles (dyeing and printing) are energy intensive industries. Steam is used as energy carrier for processing applications like dyeing and finishing in all textile industries. Hence boilers are the main fuel consumers in the textile industries. The main areas of waste heat recovery in textile dyeing are from boiler blow down flash steam, hot condensate flash steam and heat recovery from processed waste water.

14.6.1 Waste heat recovery systems for stenters

Textile stenters have two main purposes—convection drying so as to remove the moisture in the fabric and secondly to provide for fabric width control. During the previous stages of processing the fabric is subjected to length wise tension to varying degrees resulting in shrinkage in width. In the stenter, width control is achieved with the aid of a series of clips or pins mounted on a pair of endless chains.

Apart from these functions stenters are also used for the following:

1. Dry-heating process like, heat setting of synthetic fabrics and their blends.
2. Dry curing process namely, resin finishing with built-in catalysts.
3. Partial curing of pigments dyeing stenters being a major energy consumer in a textile mill offers opportunities for energy conservation.

Drying process: Drying is achieved by impinging high velocity air jets uniformly across the full width of the fabric on both sides. The air being used is heated to a temperature of about 140–150°C. The hot air is recalculated and a certain amount of air is continuously removed from the system through exhaust fans so as to avoid buildup of excessive humidity. To that extent, the system is supplemented by fresh air. The stenters located in the processing section are major consumers of steam in any textile unit. The stenters are being used for drying, stretching and finishing. The fabric enters the stenters after the pre-drying cylinders with moisture of about 60–65%. This moisture needs to be dried and vented out in the stenters.

The stenters have normally two exhaust blowers which are operating continuously venting hot air and moisture at temperatures around 100°C. At the processing plant the jigger dyeing section needs hot water at temperatures ranging from 40°C to 80°C. Presently steam is being used for supplying this heat. There is a good potential to install waste-heat recovery systems for stenter exhaust and utilise this recovered heat for dyeing machines.

Energy saving: A 15000 TPA composite mill in India has installed waste heat recovery system for stenters and has achieved energy saving of 20–30% in fuel consumption.

14.6.2 Installation of heat recovery system in merceriser machines

A treatment of cotton yarn or fabric to increase its luster and affinity for dyes. The material is immersed under tension in a cold sodium hydroxide (caustic soda) solution in warp or skein form or in the piece and is later neutralised in acid. The process causes a permanent swelling of the fibre and thus increases its luster. It is the process of treatment of cellulosic material with cold or hot caustic conditions under specific conditions to improve its appearance and physical as well as chemical properties.

Purpose of mercerising:

1. To improve the lusture.
2. To improve the strength.
3. To improve the dye uptake and moisture regain.

Mercerising process:

The mercerising involves these three subsequent steps.

1. Impregnation of the material in relaxed state, cold caustic solution of required strength and wettability.
2. Stretching while the material is still impregnated in the caustic solution.
3. Washing off the caustic soda from the material while keeping the material still in the stretch state.

Mercerising is a very important stage of textile processing. It consists of treating the fabric in a stretched condition (fabric not allowed to shrink), with 270 grams/litre caustic soda solution giving a dwell time of 50 seconds. The caustic is washed off while in the stretched stage. Residual caustic is washed with hot water using a counter current system. The counter current washing consists of a series of water baths heated with steam. The baths are inter-connected with each other. The fresh water makeup enters at one end while the spent water at about 90°C is drained off at the other end. First sent to a filter and then to pathogenic eukaryotes study section (PTHE) before draining at a temperature of 40°C. A water pump was installed to pump makeup water at a temperature of 30°C through the PTHE. The makeup water could be pre-heated to a temperature of 80°C.

Energy saving

A 15000 TPA composite mill in India has installed waste heat recovery system for mercerise and has achieved energy saving of 20–30% in terms of fuel consumption. The heat recovery system resulted in reduced steam consumption in the water baths, which in turn reduced the fuel (LSHS) input to the boiler to the tune of 122 T/year. The addition of two water pumps (for two Mercerises) increased the power consumption by a marginal 10 kW.

14.6.3   Heat recovery in bleaching range

Bleaching is a chemical treatment employed for the removal of natural colouring matter from the substrate. The source of natural colour is organic compounds with conjugated double bonds, by doing chemical bleaching the discolouration takes place by the breaking the chromophore, most likely destroying the one or more double bonds with in this conjugated system. The material appears whiter after the bleaching.

Natural fibres, i.e., cotton, wool, linen etc., are off-white in colour due to colour bodies present in the fibre. The degree of off-whiteness varies from batch-to-batch. Bleaching therefore can be defined as the destruction of these colour bodies. White is also an important market colour so the whitest white has commercial value. Yellow is a component of derived shades. For example, when yellow is mixed with blue, the shade turns green. A consistent white base fabric has real value when dyeing light to medium shades because it is much easier to reproduce shade matches on a consistent white background than on one that varies in amount of yellow.

Bleaching may be the only preparatory process or it may be used in conjunction with other treatments, e.g., desizing, scouring and mercerising. The combination of such treatments for an individual situation will depend on the rigorousness of the preparation standard and economic factors within the various options. Other chemicals will be used in addition to the bleaching agent. These serve various functions such as to activate the bleaching system, to stabilise or control the rate of activation, to give wetting and detergent action, or to sequester metallic impurities. This section gives consideration to the selection of bleaching agents and to the role of the various chemicals used in conjunction.

During the bleaching process, the fabric is treated with sodium hypo-chlorite or hydrogen peroxide. After bleaching the cloth is thoroughly washed in a series of baths.

The baths are maintained at different temperatures by direct injection of steam. The initial temperature of water required is 55°C and final temperature 85°C. The hot water is generated by direct injection of steam. The used water in the baths was being drained separately at different temperatures.

Energy saving

A 15000 TPA composite mill in India has installed waste heat recovery system for bleaching range which recovered the heat in the drain water and thus annual saving of energy was around 1.5 crore.

14.6.4   Boiler blow down heat recovery

As water evaporates in the boiler steam drum, solids present in the feed water are left behind. The suspended solids form sludge or sediments in the boiler, which degrades heat transfer. Dissolved solids promote foaming and carryover of boiler water into the steam. To reduce the levels of suspended and total dissolved solids (TDS) to acceptable limits, water is periodically discharged or blown down from the boiler. Mud or bottom blow down is usually a manual procedure done for a few seconds on intervals of several hours. It is designed to remove suspended solids that settle out of the boiler water and form a heavy sludge. Surface or skimming blow down is designed to remove the dissolved solids that concentrate near the liquid surface. Surface blow down is often a continuous process.

Minimising blow down rate can substantially reduce energy losses, as the temperature of the blow down liquid is the same as that of the steam generated in the boiler. Minimising blow down will also reduce makeup water and chemical treatment costs. Insufficient blow down may lead to carryover of boiler water into the steam, or the formation of deposits. Excessive blow down will waste energy, water and chemicals. It is necessary to control the level of concentration of the solids and this is achieved by the process of 'blowing down', where a certain volume of water is blown off and is automatically replaced by feed water-thus maintaining the optimum level of total dissolved solids in the boiler water. Blow down is necessary to protect the surfaces of the heat exchanger in the boiler. However, blow down can be a significant source of heat loss, if improperly carried out.

Concept of flash steam

Flash steam is vapour or secondary steam formed from hot condensate discharged into a lower pressure area. It is caused by excessive boiling of the condensate which contains more heat than it can hold at the lower pressure. Flash steam occupies many times the volume of water from which it forms. For example, flash steam created by hot condensate flowing from 15 psig to an atmospheric pressure will have nearly 1600 times the volume of the high pressure hot water.

Heat content in the flash steam from boiler blow down and condensate can be recovered back to pre-heat the boiler feed water and flash steam produced due to excess boiler blow down also can reduced fuel consumption rate. It is found that flash steam recovery from the boiler blow down also increases the efficiency of the boiler up to 2%. Heat recovery is used frequently to reduce energy losses that result from boiler water blow down.

14.6.5   Condensate heat recovery

Steam contains two types of energy: latent and sensible. When steam is supplied to a process application (heat exchanger, coil, etc.), the steam vapour releases the latent energy to the process fluid and condenses to a liquid condensate. The condensate retains the sensible energy the steam had. The condensate can have as much as 16% of the total energy in the steam vapour, depending on the pressure.

Figure 14.3 illustrates a typical condensate recovery system. One of highest return on investments is to return condensate to the boiler. As fuel costs continue to rise, it's imperative to focus on recovering condensate in every industrial steam operation. Returning hot condensate to the boiler makes sense for several reasons. As more condensate is returned, less makeup water is required, saving fuel, makeup water and chemicals and treatment costs. Less condensate discharged into a sewer system reduces disposal costs. Return of high purity condensate also reduces energy losses due to boiler blow down.

Flash steam recovery

Condensate is discharged through traps from a higher to a lower pressure. As a result of this drop in pressure, some of the condensate will then reevaporate into flash steam. The flash steam generated can contain up to half of the total energy of the condensate, hence flash steam recovery is an essential part of an energy efficient system.

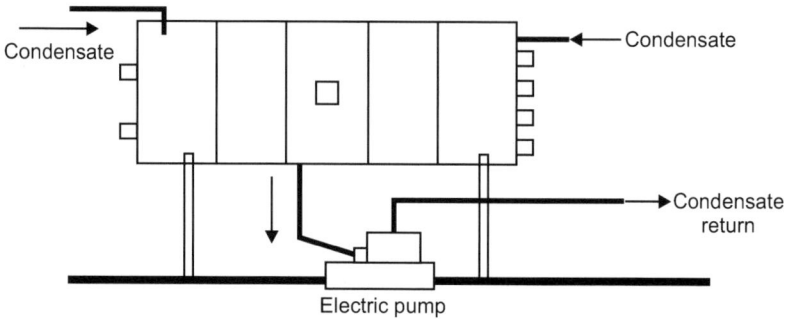

Figure 14.3: Pump and condensate receiver.

It is clear that, flash steam recovery from hot condensate enhanced boiler efficiency and it will in turn reduce the fuel consumption rate. The return of condensate represents huge potential for energy savings in the boiler house. Condensate has high heat content and approximately 1% less fuel is required for every 6°C temperature rise in the feed tank. The more the condensate recovery, the lesser will the condensate that is discharged into a sewer system be and the lower will the blow down be. This will reduce the sewer disposal costs.

Heat recovery from water

The first heat recovery option to consider is the reuse of the hot wastewater. In this way, water, residual chemicals as well as energy are recovered. In textile dyeing and finishing, operations involving acrylic fibres or wool where colourants are exhausted, wastewater reuse is possible. Similarly, wastewater from rinsing operations can make up new baths, for instance, for scouring. Dyeing and finishing specialists claim that wastewater from light shade operations can be reutilised up to 20 times.

Cooling water recovery

Cooling of baths is a common operation. The utilisation of cooling water, that is, of a stream of cold water to absorb heat from the hot bath, can also be considered as a heat recovery process. Subsequently, cooling water is collected and reutilised, thus, recovering heat and water. Under the most favourable conditions, cooling water recovery has been reported to have a payback period of 12 months.

Heat recovery from wastewater

Batch or non-continuous processing is common in textile dyeing plants. Thus, a large volume of wastewater is available intermittently from several machines at different locations in the plant. If wastewater can neither be reused nor can its heat be recovered locally, the feasibility of installing a centralised heat recovery system should be investigated. Figure 14.4 shows a typical setup for centralised wastewater recovery.

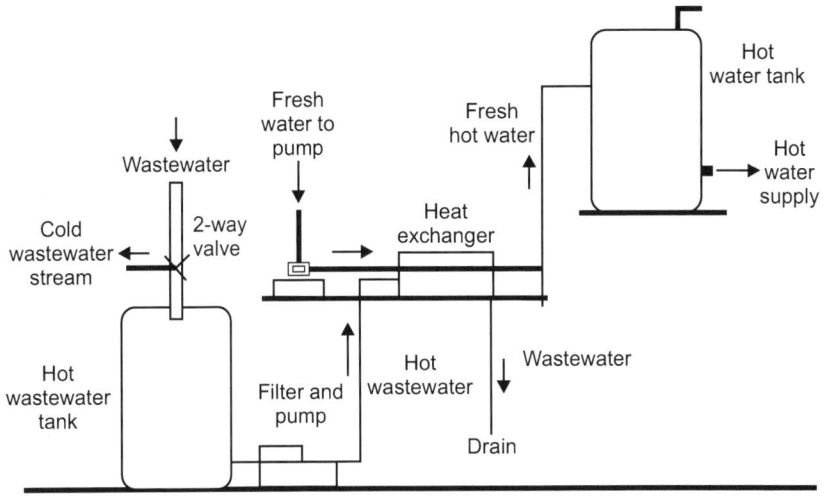

Figure 14.4: Centralised waste water heat recovery.

Equipment such as washing-, mercerising- and bleaching-machines often operate continuously for long hours, requiring a large volume of hot water and produce an equal volume of hot waste water simultaneously. This can be done by incorporation of heat exchangers on such textile machines with the purpose of heating up the incoming cold-water stream with hot wastewater leaving the machine. This water can be properly utilised for pre-heating the boiler feed water or dyeing purpose, it can save energy as well as water. Hot wastewater, produced in textile dyeing plants, can be a significant source of heat energy. In many instances, this valuable resource is discharged to wastewater treatment facilities without employing the heat it acquired during processing. Most of the heat contained in the wastewater stream can be reclaimed and utilised, while providing significant cost reductions with attractive payback periods. Through utilisation of proper wastewater heat recovery system, reclaimed heat from the wastewater discharge can pre-heat incoming process water; thereby saving fuel costs, while enhancing the environment through the removal of thermal pollution.

Wastewater heat exchangers

Shell-and-tube and plate heat exchangers have been used with mixed results in heat recovery in industry. The latter is generally preferred to the former because its heat transfer efficiency is higher by 3 to 5 times. Plate heat exchangers require less capital investment than shell-and-tube heat exchangers made of materials like high-grade stainless steel, essential to resist corrosion due to textile waste-water. However, both are affected by fouling.

The heat exchanger has the following characteristics:

1. It can work without filters and is self-cleaning as a result of the rotation of the scroll preventing the accumulation of fibres, particulates and other foulants.

2. Heat exchanger efficiency is enhanced as a result of agitation caused both inside the scroll for the fresh water stream and outside the scroll for the wastewater stream.

3. The pressure drop due to wastewater flow is lower than in the case of other heat exchangers and negligible power is needed to rotate the scroll at low speed.

4. Less space is required for the heat exchanger than for a shell-and-tube heat exchanger for the same heat recovery duty. Unless the heat exchanger is used on continuously operating machines, storage tanks will also be required.

The spiral flow or rotary heat exchanger had a payback of less than 12 months if applied on continuously operating machines with high wastewater flow. A survey in 2010 in the U.K. showed that 26 % of textile dyeing and finishing plants proposed to adopt the latter heat exchanger. Such units have been used for decades to date in different plants.

14.7   Cost effectiveness in textile processing

A few important guidelines for various cost saving measures by process modifications and adopting new concepts based on modern technological changes are given.

14.7.1   Energy and water conservation

In every country, the growing needs of energy in various fields of activities have led to the necessity of finding out ways and means to avoid wastage and conserve the energy. Water is expensive to buy, treat and dispose. Textile industries are consuming large quantity of water. Major portion of water is used for wet processing of textile (60 to 70 %). Keeping this in mind there is acute need of energy and water conservation programmed to implement as earlier as possible.

Energy conservation

Energy is one of the most important ingredients in any industrial activity. However, its availability is not infinite. Global energy crisis, as well as high cost of fuels resulted in more activities to conserve energy to maximum extent.

The textile industry retains a record of the lowest efficiency in energy utilisation and is one of the major energy consuming industries. About 34% of energy is consumed in spinning, 23% in weaving, 38% in chemical wet processing and another 5% for miscellaneous purposes. Power dominates consumption pattern in spinning and weaving, while thermal energy is major for chemical wet processing.

14.7.2 Preventive energy conservation finding-energy losses and solutions of them

Following are the major sources from where energy is lost in various forms:

Energy loss through hot vapours and hot water discharge

A large amount of thermal energy, either in the form of hot exhaust gases or hot vapours is ejected into the atmosphere or down the drain as waste from various processing machinery in the textile mills. The recovery of such waste for reuse is estimated to effect 5 to 7% saving in the cost. For example:

1. Flue gases in plain and thermic fluid boilers.
2. Liquor, water and their vapours in the kiers, agers, jiggers, jet dyeing and beam dyeing machines.
3. Exhaust gases from the stenters.

Major sources of hot wastewater discharge are desizing washes, scouring, mercerisation, dying and steam condensate wastage. Most of the wastewater streams are discharged at the temperature of 60–70°C.

Heat energy can be recovered from the hot wastewater streams originating from different processes by installing heat exchanger between the wastewater and fresh water to be used for different process. The capacity of the heat exchanger will vary with the discharge of the machine. The temperature rose of the cold water would be in the range 40–50°C. This measure will reduce steam consumption for heating the baths. Different types of heat exchangers are available and its choice depends on number of factors. Steam condensate discharges are hot and clean water streams. These streams can be used as boiler feed water or for preparation of dye baths. These options not only reduce water consumption and wastewater quantities, but also results into substantial energy savings.

Energy loss through leakages and improper maintenance

In most of the textile industries it was found that there are number of pipelines and equipment from where steam, steam condensate and hot water is lost through leakages. It is difficult to asses the quality of leakages but obvious that leakage of hot water and steam results in substantial energy loss. Generally the condition of piping and insulation is not up to the standard due to the fact that preventive maintenance is not being given due consideration. One of the reasons for this negligence may be due to production load in which machine shut down for repair is difficult.

Steam usage is generally not optimum, reasons for excess usage and wastage of steam are the unnecessary supply of steam to the bath even after attaining required temperature. Steam traps are generally malfunctioning, resulting into escape of steam along with steam condensate. At most of the places, out of order steam traps are disconnected, rather being repaired or replaced. Corroded pipes and valves, as a result of improper maintenance, also contribute in steam and hot water loss. Steam control valves are generally not found in the machines and old machines are not equipped with energy controllers.

Preventive maintenance should be given due consideration. Routine maintenance schedule should be maintained and followed properly. Workers should be aware of the fact that substantial money is lost through leakages and improper maintenance practices.

Energy loss through lack of insulation on pipelines and machines

Most of steam, steam condensate and hot water carrying pipelines are not equipped with proper insulation. Insulation found to be eroded at various places because of improper maintenance and upkeep. Machines conducting reactions, washing and drying at hot states especially desizing, bleaching, jiggers and dryers are mostly not insulated, that cause heat dissipation to ambient air. The quantity of heat dissipation is the function of the temperature difference between inside hot machine and outside cold air and the surface area of the machine. However, proper insulation provides resistance to convectional heat transfer with the advantage of less steam and fuel consumption in heating contents up to the required temperature.

A saving of the order of 5–10% of the energy consumed for steam production can result from this measure. Further this measure will improve the occupational atmosphere, especially during the hot weather and also increase the safety of the workers, due to covering of the otherwise hot surfaces.

Energy loss through flue gases and hot air

Boiler flue gases contain substantial heat energy. This energy can be utilised to pre-heat the boiler feed water through economiser but at present in most of the industries it is not being utilised. Some industries installed economiser but could not last, may be because of improper material of construction or faulty design. Hot air, from different dryers, is also wasted. The thermal energy, present in the boiler off-gas stream, can be used to pre-heat the boiler feed water. Savings in boiler fuel consumption can amount to about 5%. Economiser can be boiler stack in which hot flue gases will pre-heat boiler feed water.

Energy loss through singeing machine

Singeing operation is employed to destroy singes and tufts on the surface of the fabric, by its direct exposure to the flame, for a very short time. Fabrics of varying widths are processed in the textile industry. Generally a single multiple port full-width burner is provided in the singeing machine, which results into unnecessary wastage of energy, when fabric of width shorter that of the singeing flame is being processed.

The arrangement required would be the replacement of the single full width burner with a series of burners of shorter widths placed in one row, each with separate fuel gas supply. The number of burners to be fired, during singeing operation, would be according to the width of the fabric.

Use of non-conventional energy sources

The different alternative renewable resources of energy are biomass, geothermal energy, tidal energy, wind energy and solar energy. Out of these energy sources, solar energy is abundant and is inexhaustible, in fact, fossil fuel, viz., coal, oil and natural gas owe to their origin to these energy sources. India's geographical location favours unlimited and uninterrupted trapping of solar energy and it is the desirable energy available in the environment.

The plants serve as the most abundant renewable raw material in nature for production of biogas, as they are rich in carbohydrates. The gas can be produced and consumed at the place of production and hence cost of transportation of raw material and gaseous product is eliminated. The technology is simple and easy to operate, with virtually very little maintenance cost. There will not be any problem of air pollution. In short, nothing is wasted and there is no effluent.

Steam generation and its effective distribution

Steam has become the first source of heat in dye house. The only way to absorb all the heat from the steam is to inject source of heat in dye bath. Heat exchanger of coil, tube or plate type, suitably closed by a steam trap to permit the drainage of condensed steam, are now a days installed in a machine itself. Most of the energy goes for generation of steam and production of power. The efficiency of conversation of fuel into steam is of the order of 75% and that of electricity about 30%. The steam generated from the boiler has to be distributed through proper insulated pipelines wherever required. The loss due to the leakage have to be ventilated properly to avoid the condensing of the steam on the roofs and the machines like jigger and winches may also be covered with covers.

Drying

Drying is a thermal energy consuming process required at different stages in wet processing.

Drying operation is done by different techniques.

1. Cylinder drying can be made more effectively by use of steel cylinders and reducing water contents by vacuum extraction.
2. Hot air stenter, if used by preventing fresh air leakage, running exhaust only when necessary, keeping circulation of air effective and avoiding idle running can save energy.
3. Infrared and RF-drying are also effective due to their specific effects.
4. Microwaves are more energy efficient saving in both capital and operating cost. Microwaves affect polar molecules (such as water) and in a textile material the substrate is generally non polar, hence water is evaporated without affecting the substrate by IMS (Industrial Micro-wave System).

Process modification

Bleaching and finishing:

1. Combined preparatory processes save time as well as energy.
2. If the fabric is pretreated by subjecting grey one to steam purging operation, wettability improves, which can be followed by single step desizing, scouring and bleaching.
3. Explore scope for wet-on-wet mercerising and wet-on-wet finishing by using vacuum extractor.
4. Run two or more ends either side-by-side or superimposed on machine like shearing, cropping, sizing, chainless merceriser, cylinder drying rages and calendars, etc.
5. Combined drying and heat setting or optical brightening and heat setting.

Dyeing and finishing:

1. Dyeing of blends in single bath.
2. Combined dyeing and finishing process.
3. Solvent assisted dyeing, supercritical carbon dioxide as a medium of dyeing.
4. Ultrasonic and UV energy for dyeing.
5. Cold pad bath operations wherever possible.
6. Eliminating intermediate drying operation between beam dyeing of polyester and jigger dyeing of cellulose in blends.
7. Resort to drying cum curing of pigment prints.

Thermal and electrical conservation in textile industry

15.1   Introduction

Heat generated through the combustion of fuel is utilised to generate steam from water in a boiler and to heat thermic fluid in a thermic fluid heater. Heat of the steam and the thermic fluid is utilised in various textile processors. Ever escalating fuel cost has increased the cost of heat energy to a considerable extent which is one of the important factors contributing to the higher processing cost and hence the production cost of textiles. Higher production cost thin profit margins, tough global competition collectively have posed a serious problem for survival of textile industry. Therefore, every possible step has to be taken to lower the production cost. Heat economy, i.e., conservation of thermal energy is a major option in this regard. Heat economy is nothing but efficient generation and utilisation of heat energy. Scope for heat economy, i.e., conservation of thermal energy exists during generation and utilisation of thermal energy, recovery of waste heat, etc.

15.2   Thermal energy conservation in textile industry

Textile industry consumes substantial quantity of thermal as well as electrical energy. Thermal energy consumption accounts for about 70% of the total energy consumption in a composite textile mill. A steep hike in the price of fuel during the last few years has escalated the cost of thermal energy many folds. Besides, erratic and inadequate availability of fuel in general and liquid fuels in particular, has made the situation alarming. Therefore, in the wake of such a critical situation, thermal energy conservation is no more a choice, but has become a bare necessity. Thermal energy conservation is nothing but optimisation of the use of thermal energy. Areas to minimise the consumption of thermal energy have to be identified first and then suitable action taken to reduce thermal energy levels. This section deals with the measures for thermal energy conservation involving low capital investments based on the findings made and recommendations suggested to the textile mills by the Energy and Engineering Maintenance Audit (EEMA).

Intensive studies for thermal energy conservation has been conducted in more than 80 textile mills spread all over India during the past one decade. On an average, a mill can save around 14–18% of the thermal energy cost by implementation of simple, less capital intensive conservation measures.

15.2.1   Storage and handling of fuel

Coal

1. Make the ground of the coal storage yard firm and hard to avoid carpet loss of coal. Soft and uneven ground can cause carpet loss as high as 2–3%.
2. Stack different qualities of coal separately and blend them before crushing to get uniform quality for feeding into boilers.
3. Construct coal storage bays with proper slope and with arrangements for drainage and coal stock measurement.
4. Make arrangement for sprinkling of water on coal stacks to minimise wind loss and to avoid spontaneous combustion.
5. Ensure proper distance between coal storage yard and ash dumping ground to avoid inadvertent mixing of coal and ash.
6. Weigh the coal before use.
7. Reconcile coal periodically. Coal receipts, consumption, stock should be reconciled to keep a check on coal losses.
8. Break coal lumps to smaller pieces for their efficient combustion into the following size range:

   For hand fired boilers   —   40–50 mm
   For stoker fired boilers   —   15–20 mm
9. Ensure wetting of coal (broken) for optimum combustion of coal fines (it has been observed that if the water sprinkling is restricted to 1–1.5% for every 10% of fines, the gains due to fines burning would exceed the loss due to latent heat of evaporation of moisture added).

Fuel oil

1. Provide unloading platform near the oil storage tanks for proper unloading of fuel oil.
2. Drain service tanks daily and storage tanks once a week to remove sludge and water from fuel oil. Clean service tanks quarterly and storage tanks yearly. Removal of sludge and water would help in better atomisation and efficient combustion of fuel.
3. Provide proper system of duplex filter at various points in the storage and handling system. This would help in efficient removal of foreign suspended matter from oil thereby reducing wear and tear of burner nozzles.
4. Restrict electrical heating of oil to initial heating only, after which switch over to steam heating as the former is costlier than the latter.

5. Pre-heat the fuel oil to the following temperatures to ensure proper atomisation.

   Furnace oil — 105–115°C

   Low sulphur heavy stock — 90–100°C

6. Ensure proper lagging of fuel oil lines, tanks, etc., to minimise losses of thermal energy.

7. Take proper care to avoid wastage of oil through leakages, etc., during unloading and handling of fuel oil.

15.2.2 Generation of thermal energy, steam generation and thermic fluid heating

1. Maintain the auto controls of boilers in working order to ensure safe and efficient working of boilers.

2. Check CO_2/O_2 (%) in flue gases for every shift. Any deviation from the target value (for the given fuel and type of burner) should be rectified immediately as follows:

   (a) In the case of oil fired boilers: Adjustment of dampers, maintaining required fuel oil pre-heat temperature and pressure, cleaning of burner nozzle, etc.

   (b) In the case of coal fired boilers: Adjustment of dampers, balancing the draught, employing proper technique of coal firing/ash removal, maintaining thickness of coal bed around 6 inch on the grate of boiler, plugging the cracks in the brick work, etc.

3. Monitor flue gas temperature and rectify deviations from target value by cleaning heat transfer surfaces, stopping air infiltration, etc.

4. After maintaining the set values of flue gas temperature and CO_2/O_2 (%) in the flue gases, the target levels of thermal efficiency could be attained if proper care is taken to minimise other losses due to incomplete combustion of coal, clinker formation, radiation, etc.

5. Ensure proper coal size and wetting of coal to reduce losses due to incomplete combustion.

6. Select the coal properly and employ proper firing methods to avoid loss due to clinker formation.

7. Insulate brick work properly to reduce loss due to radiation.

8. Load the boilers to around 80–85% of their rated capacity for optimum performance.

9. Avoid banking losses of boiler, operate only the required number of boilers as per the steam demand.

10. Avoid frequent start/stop of the boilers. Boilers working on frequent start/stop operation lose their efficiency by 2–3%. If the boiler capacity is in excess of the demand considerably, curtail fuel input to the boilers.

11. Limit the boiler steam pressure depending upon the requirement of the machines making due allowance for the pressure drop in the steam distribution system.

12. Adjust boiler blow down keeping in view the total dissolved solids levels prescribed for different types of boilers. This will minimise thermal energy losses due to boiler blow down.

13. Maintain proper feed water treatment to minimise scale formation on heat transfer surfaces. A scale deposit of even 1.6 mm thickness reduces heat transfer efficiency by 10%

14. Restrict the heating of thermic fluid depending upon the process/machine requirement. This will reduce flue gas losses as well as the radiation losses.

15. Monitor the flue gas temperature of the thermic fluid heater and if abrupt rise in temperature is noticed, then check for short circulating of the flue gases.

15.2.3   Utilisation of thermal energy

1. Install steam mains and branch lines of proper size to avoid any pressure drop and thermal energy loss.

2. Recommended steam velocity for designing the steam lines is as follows:
   Main lines   —   24–30 m^3/sec
   Branch lines   —   15 m^3/sec

3. Install boiler house close to the steam usage points consuming steam at higher rates to minimise the loss of thermal energy.

4. Dismantle redundant steam/condensate/thermic fluid lines to avoid unnecessary loss of thermal energy.

5. Ensure supply of dry steam to the machines by employing proper condensate removal arrangement for main and branch lines.

6. Provide/set right pressure reducing valves in the steam lines of machines in order to obtain steam at the required lower pressure. This will reduce the consumption of steam.

7. Avoid throttling of valve to control steam pressure as it is not only an unsatisfactory method to control the down stream pressure, but it also damages the valve.

8. Use low pressure steam (1–1.5 kg/cm^2 pressure) for direct heating of the liquids.

9. Provide pressure gauge in steam supply line for monitoring steam pressure.

10. Provide steam line of larger size after the pressure reducing valve than that before so as to accommodate the increase in volume due to pressure reduction.

11. Provide air vent in the condensate removal lines of machines for complete removal of air so that rate of heat transfer within the machine is improved.

12. Provide traps of proper type and size and install them properly for effective condensate removal.

15.3   Electrical energy conservation in textile industry

Ever increasing cost of energy has made an adverse impact on the profitability of the textile mills. For a composite mill, the requirement of both thermal and electrical energy is large. Thermal energy is consumed mainly in the chemical processing department, whereas the consumption of electrical energy is more in the spinning and weaving departments.

A decade ago, cost of electrical power was 4–5% of the total cost of production; however, at present, it is about 12–15%. Besides, inadequate availability of power is another constraint. It has, therefore, become imperative to pay more attention to the effective utilisation of electrical energy. The measures to conserve energy have to be spread over different time frames, viz., long term, medium term and short term.

The long and medium term measures involve considerable capital expenditure and need careful scrutiny. However, what deserves immediate implementation in the textile mills is the energy saving measures which can be called as house-keeping measures, capable of yielding immediate returns without much capital investment.

15.4   Energy conservation techniques in textile industry

15.4.1   EC techniques in transformers

1. Optimisation of loading of transformer:
   (a) By proper location of transformer preferably close to the load center, considering other features like centralised control, operational flexibility, etc. This will bring down the distribution loss in cables.
   (b) Maintaining maximum efficiency to occur at 38% loading [as recommended by Rural Electrification Corporation Limited (REC)], the overall efficiency of transformer can be increased and its losses can be reduced.

(c) Under fluctuating load condition, more than one transformer is used in parallel operation of transformers to share the load and can be operated close to the maximum efficiency range.

2. By improvisation in design and material of transformer:
    (a) To reduce load losses in transformer, use thicker conductors so that resistance of conductor reduces and load loss also reduces.
    (b) To reduce core losses use superior quality or improved grades of Cold Rolled Grain Oriented (CRGO) laminations.

3. Replacing by energy efficient transformers:
    (a) By using energy efficient transformers, efficiency improves to 95–97%.
    (b) By using amorphous transformers, efficiency improves to 97–98.5%.
    (c) By using epoxy resin cast/encapsulated dry type transformer, efficiency improves to 93–97%.

Transformer losses can be divided into two main components: No-load losses and Load losses. These types of losses are common to all types of transformers, regardless of transformer application or power rating.

There are, however, two other types of losses; extra losses created by harmonics and losses which may apply particularly to larger transformers– cooling or auxiliary losses, caused by the use of cooling equipment like fans and pumps.

No-load losses

These losses occur in the transformer core whenever the transformer is energised (even when the secondary circuit is open). They are also called iron losses or core losses and are constant.

They are composed of:

1. Hysteresis losses, caused by the frictional movement of magnetic domains in the core laminations being magnetised and demagnetised by alternation of the magnetic field. These losses depend on the type of material used to build a core. Silicon steel has much lower hysteresis than normal steel but amorphous metal has much better performance than silicon steel. Nowadays hysteresis losses can be reduced by material processing such as cold rolling, laser treatment or grain orientation. Hysteresis losses are usually responsible for more than a half of total no-load losses (~50% to ~70%).

2. Eddy current losses, caused by varying magnetic fields inducing eddy currents in the laminations and thus generating heat. These losses can be reduced by building the core from thin laminated sheets insulated from each other by a thin varnish layer to reduce eddy currents. Eddy

current losses nowadays usually account for 30–50% of total no-load losses. When assessing efforts in improving distribution transformer efficiency, the biggest progress has been achieved in reduction of these losses.

3. There are also marginal stray and dielectric losses which occur in the transformer core, accounting usually for no more than 1% of total no-load losses.

Load losses

These losses are commonly called copper losses or short circuit losses. Load losses vary according to the transformer loading.

They are composed of:

1. Ohmic heat loss: Ohmic heat loss, sometimes referred to as copper loss, since this resistive component of load loss dominates. This loss occurs in transformer windings and is caused by the resistance of the conductor. The magnitude of these losses increases with the square of the load current and is proportional to the resistance of the winding. It can be reduced by increasing the cross sectional area of conductor or by reducing the winding length. Using copper as the conductor maintains the balance between weight, size, cost and resistance; adding an additional amount to increase conductor diameter, consistent with other design constraints, reduces losses.

2. Conductor eddy current losses: Eddy currents, due to magnetic fields caused by alternating current, also occur in the windings. Reducing the cross-section of the conductor reduces eddy currents, so stranded conductors are used to achieve the required low resistance while controlling eddy current loss.

   Effectively, this means that the 'winding' is made up of a number of parallel windings. Since each of these windings would experience a slightly different flux, the voltage developed by each would be slightly different and connecting the ends would result in circulating currents which would contribute to loss.

   This is avoided by the use of Continuously Transposed Conductor (CTC), in which the strands are frequently transposed to average the flux differences and equalise the voltage.

Auxiliary losses

These losses are caused by using energy to run cooling fans or pumps which help to cool larger transformers.

Extra losses due to harmonics and reactive power

This category of losses includes those extra losses which are caused by reactive power and harmonics. The reactive component of the load current generates a real loss even though it makes no contribution to useful load power. Low power factor loads should be avoided to reduce losses related to reactive power. Power losses due to eddy currents depend on the square of frequency so the presence of harmonic frequencies which are higher than normal 50 Hz frequency cause extra losses in the core and winding.

Extra losses due to harmonics

Non-linear loads, such as power electronic devices, such as variable speed drives on motor systems, computers, UPS systems, TV sets and compact fluorescent lamps, cause harmonic currents on the network. Harmonic voltages are generated in the impedance of the network by the harmonic load currents. Harmonics increase both load and no-load losses due to increased skin effect, eddy current, stray and hysteresis losses. The most important of these losses is that due to eddy current losses in the winding; it can be very large and consequently most calculation models ignore the other harmonic induced losses.

The precise impact of a harmonic current on load loss depends on the harmonic frequency and the way the transformer is designed.

In general, the eddy current loss increases by the square of the frequency and the square of the load current. So, if the load current contained 20% fifth harmonic, the eddy current loss due to the harmonic current component would be $5 \times 5 \times 0.2 \times 0.2$ multiplied by the eddy current loss at the fundamental frequency–meaning that the eddy current loss would have doubled.

In a transformer that is heavily loaded with harmonic currents, the excess loss can cause high temperature at some locations in the windings. This can seriously reduce the life span of the transformer and even cause immediate damage and sometimes fire.

Power factor and load management

1. Improve power factor above 0.96 by: (i) selection of proper size of motor for a drive and (ii) installation of capacitors of low dielectric losses such as of polypropylene or any other mixed dielectric type.
2. Balance capacitors as per the load.
3. Install capacitors nearer to the point of consumption.
4. Stagger timings for working of machines and recess.
5. Avoid idle running of machines.
6. Inform the power house before switching on heavy loads.

7. Maintain a high load factor.
8. Stagger starting and stopping of high horsepower motors.
9. Incorporate a warning system in the maximum demand indicator so as to enable immediate action.
10. Install power factor sensing relay for automatic switching of capacitors depending upon the power factor of the system.

15.4.2   Energy efficient transformers

Most energy loss in dry-type transformers occurs through heat or vibration from the core. The new high-efficiency transformers minimise these losses. The conventional transformer is made up of a silicon alloyed iron (grain oriented) core. The iron loss of any transformer depends on the type of core used in the transformer. However the latest technology is to use amorphous material - a metallic glass alloy for the core. The expected reduction in energy loss over conventional (Si Fe core) transformers is roughly around 70%, which is quite significant. By using an amorphous core- with unique physical and magnetic properties- these new type of transformers have increased efficiencies even at low loads–98.5% efficiency at 35% load.

Electrical distribution transformers made with amorphous metal cores provide excellent opportunity to conserve energy right from the installation. Though these transformers are a little costlier than conventional iron core transformers, the overall benefit towards energy savings will compensate for the higher initial investment. At present amorphous metal core transformers are available up to 1600 kVA.

15.4.3   Manufacturing and wet processing machinery

1. Check loading of motors and if the load is below 60%, replace the motor by a lower capacity one.
2. Replace inefficient motors with more efficient ones.
3. Prefer individual drive to group drive for better control and constant speed.
4. Interlock auxiliary drive motors with the main motor of the machine.
5. Provide stop motion for clutch type looms to avoid unnecessary power consumption due to idle running.
6. Ensure optimum utilisation of the machine capacity.
7. Pay attention to regular upkeep and maintenance of the machines.
8. Replace AC drive with DC motor and rectifier dimmerstat control drive for machines requiring varying speed for optimum power consumption.

9. Replace electrical heating with steam or thermic fluid heating.
10. Avoid re-processing.
11. Consider process sequence modifications in such a way that a few unit operations get eliminated.

Humidification and ventilation systems

1. Control the air flow by changing pulleys.
2. Run only the required number of air-supply fans.
3. Stop use of water spray pump in the monsoon season.
4. Provide false ceiling to reduce heat load.
5. Replace oversized motors.
6. Replace inefficient motors.
7. Reduce air infiltration.
8. Provide measures for cooling of roof to minimise heat load.
9. Maintain optimum RH and dry bulb temperature inside the sections/departments.
10. Control flow of return/exhaust air.
11. Attend to the regular maintenance of the plant/systems.
12. Check cooling effectiveness efficiency of air washer plants.

Compressed air system

1. Ensure proper selection of compressor to perform a specific duty.
2. Prefer centrailsed system vis-a-vis the decentralised one.
3. Select properly the site for installation of the compressor.
4. Discourage use of compressed air for human comforts and floor cleanings, etc.
5. Provide air flow guns for cleaning to avoid wastages.
6. Evaluate compressor efficiency periodically.
7. Provide automatic air-traps and other accessories.
8. Test periodically the air distribution system for leakages.
9. Provide pressure reducing valves.
10. Attend to the regular maintenance of the plants/systems.

15.5 Diesel generator (DG)

DG set is a combination of a diesel engine and an alternator. Diesel engine is the prime mover which drives an alternator to produce electrical energy. In the diesel engine, air is drawn into the cylinder and is compressed to a high ratio (14:1–25:1). A metered quantity of diesel fuel is then injected into the cylinder which ignites spontaneously because of the high temperature. Hence, the diesel engine is also known as compression ignition (CI) engine. DG set can be classified according to cycle type as: two stroke and four stroke. However, the bulk of IC engines use the four stroke cycle. Types of fuel or energy used in DG sets are furnace oil and diesel.

15.5.1   Design and operation

A diesel generating set should be considered as a system since its successful operation depends on the well-matched performance of the components, namely:

1. The diesel engine and its accessories.
2. The AC generator.
3. The control systems and switchgear.
4. The foundation and power house civil works.
5. The connected load with its own components like heating, motor drives, lighting, etc.

It is necessary to select the components with highest efficiency and operate them at their optimum efficiency levels to conserve energy in this system. Various components of DG set are shown in Fig. 15.1. To make a decision on the type of engine, which is most suitable for a specific application, several factors need to be considered. The two most important factors are power and speed of the engine. The power requirement is determined by the maximum load. The engine power rating should be 10–20% more than the power demand by the end use.

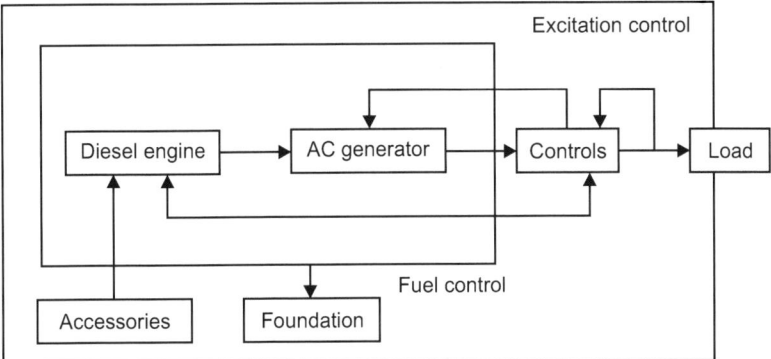

Figure 15.1: Diagram of DG set components.

This prevents overloading the machine by absorbing extra load during starting of motors or switching of few types of lighting systems or when wear and tear on the equipment pushes up its power consumption.

An engine will operate over a range of speeds, with diesel engines typically running at lower speeds (1300–3000 rpm). Speed is measured at the output shaft and given in revolutions per minute (rpm). There will be an optimum speed at which fuel efficiency will be greatest. To determine the speed requirement of an engine, one has to again look at the requirement of the load. For some applications, the speed of the engine is not critical; but for other applications such as a generator, it is important to get a good speed match. If a good match can be obtained, direct coupling of engine and generator is possible; if not, then some form of gearing will be necessary - a gearbox or belt system, which will add to the cost and reduce the efficiency.

There are various other factors that have to be considered, when choosing a diesel engine for a given application. These include cooling system, abnormal environmental conditions (dust, dirt, etc.), fuel quality, speed governing (fixed or variable speed), poor maintenance, control system, starting equipment, drive type, ambient temperature, altitude, humidity, etc. Suppliers or manufacturers literature will specify the required information when purchasing an engine.

The efficiency of an engine depends on various factors, for example, load factor (percentage of full load), engine size and engine type. With the steady development of the diesel engine, the specific fuel consumption can come down. With the arrival of modern high efficiency turbochargers, it is possible to use an exhaust gas driven turbine generator to further increase the engine rated output.

The net result would be lower fuel consumption per kWh and further increase in overall thermal efficiency. The diesel engine is able to burn the poorest quality fuel oils, unlike gas turbine, which is able to do so with only costly fuel treatment equipment.

Diesel generator (DG) set selection and installation factors

1. If a DG set is required for 100% standby, then the entire connected load in HP/kVA should be added. After finding out the diversity factor (demand/connected load), the correct capacity of a DG set can be found out.

2. For an existing installation, record the current, voltage and power factor reading at the main bus-bar of the system at every half-an-hour interval for a period of 2–3 days; and during this period, the factory should be conducting its normal operations. The non-essential loads should be switched off to find the realistic current taken for running essential equipment. This will give a fair idea about the current taken from which the rating of the set can be calculated.

3. For a new installation, an approximate method of estimating the capacity of a DG set is to add full load currents of all the proposed loads to be run

in DG set. Then, applying a diversity factor depending on the industry, process involved and guidelines obtained from other similar units, correct capacity can be arrived at.

Unbalanced load effects: It is always recommended to have the load as much balanced as possible, since unbalanced loads can cause heating of the alternator, which may result in unbalanced output voltages. The maximum unbalanced load between phases should not exceed 10% of the capacity of the DG sets.

Load pattern: In many cases, the load will not be constant throughout the day. If there is substantial variation in load, then consideration should be given for parallel operation of DG sets. In such a situation, additional DG sets are to be switched on when load increases. The typical case may be an establishment demanding substantially different powers in first, second and third shifts. By parallel operation, DG sets can be run at optimum operating points or near about, for optimum fuel consumption and additionally, flexibility is built into the system. This scheme can also be applied where loads can be segregated as critical and non-critical loads to provide standby power to critical load in the captive power system.

Energy performance assessment of DG sets

Routine energy efficiency assessment of DG sets involves following typical steps:

1. Ensure reliability of all instruments used for trial.
2. Collect technical literature, characteristics and specifications of the plant.
3. Conduct a 2 hour trial on the DG set, ensuring a steady load, wherein the following measurements are logged at 15 minutes intervals.
   (a) Fuel consumption (by dip level or by flow meter).
   (b) Amps, volts, PF, kW, kWh.
   (c) Intake air temperature, Relative Humidity (RH).
   (d) Intake cooling water temperature.
   (e) Cylinder-wise exhaust temperature (as an indication of engine loading).
   (f) Turbocharger rpm (as an indication of loading on engine).
   (g) Charge air pressure (as an indication of engine loading).
   (h) Cooling water temperature before and after charge air cooler (as an indication of cooler performance).
   (i) Stack gas temperature before and after turbocharger (as an indication of turbocharger performance).
4. The fuel oil/diesel analysis is referred to from an oil company data.

Energy saving measures of DG sets

The following options will ensure that your diesel genset is operating at best efficiency and you can tap potential energy savings:

1. Ensure steady load conditions on the DG set, and provide cold, dust free air at intake (use of air washers for large sets, in case of dry, hot weather, can be considered).

2. Improve air filtration.

3. Ensure fuel oil storage, handling and preparation as per manufacturers' guidelines/oil company data.

4. Consider fuel oil additives in case they benefit fuel oil properties for DG set usage.

5. Calibrate fuel injection pumps frequently.

6. Ensure compliance with maintenance checklist.

7. Ensure steady load conditions, avoiding fluctuations, imbalance in phases, harmonic loads.

8. In case of a base load operation, consider waste heat recovery system adoption for steam generation or refrigeration chiller unit incorporation. Even the jacket cooling water is amenable for heat recovery.

9. In terms of fuel cost economy, consider partial use of biomass gas for generation. Ensure tar removal from the gas for improving availability of the engine in the long run.

10. Consider parallel operation among the DG sets for improved loading and fuel economy thereof.

11. Carry out regular field trials to monitor DG set performance and maintenance planning as per requirements.

Energy efficient motors, gears, fans and compressors

16.1   Introduction

Electric motor systems account for about 60% of global industrial electricity consumption. Electric motors drive both core industrial processes, like presses or rolls and auxiliary systems, like compressed air generation, ventilation or water pumping. They are utilised throughout all industrial branches, though the main applications vary. Studies showed a high potential for energy efficiency improvement in motor systems in developing as well as in developed countries.

Particularly, system optimisation approaches that consider the whole motor system's efficiency show great potential. Many of the energy efficiency investments show payback times of only a few years. Still, market failures and barriers like lack of capital, higher initial costs, lack of attention by plant managers and principal agent dilemmas hamper the investment in energy efficient motor systems.

16.2   Monitoring motor failures

Present day continuous process plants demand high reliability of rotating equipments, which are mostly driven by electrical motors. Forced outages of motors could lead to production loss besides loss of energy during start up and shutdown, motor repairs and man hours lost.

Motors are designed and manufactured as per the standards laid down by various institutes and manufacturers to give trouble free operation. Series of inspections and tests are carried out during every stage of manufacture before the motors are certified and released for use. A study of more than 50,000 LT induction motor failures show that the actual life achieved is short and motor failures are more due to insulation and bearing failures. Other causes for failure are hot running of motors, over loading and jamming, corrosive fluid and water entry, single phasing. If these are attended to in time, a number of motor failures and unforeseen stoppages could be avoided.

16.2.1   Voltage variations

Insulation failure could be due to either unbalanced terminal voltage, wide variation in supply voltage compared to permissible limits given by the manufacturer or insufficient cooling. There could be unbalanced currents in the motor accompanied by a decrease in available torque and efficiency and increase in slip, vibration, noise and motor heat losses. In the phase with the highest current, the percentage increase in temperature rise will be approximately two times the square of the percentage voltage inbalance.

International specification of induction motor in U.S. are IEEE 112-B and Germany IEC 34-2 and Indian specification are IS: 13529. These specifications give the effect of unbalanced voltage on the performance of 3 phase induction motors. In case the inbalance voltage is high, motors should be derated to prevent the failure.

16.2.2   Poor insulation resistance

The insulation resistance of the windings should be tested periodically during service and maintained as per IS: 900 and BIS specification code of practice for installation and maintenance of induction motors. Insulation resistance should also be checked for standby motors before starting, during the rainy season. In case of weak insulation, resistance becomes a regular feature, the windings should be given a coat of good insulating varnish after the machine has been dried out. Standby motors should be run alternatively to avoid unexpected stoppages.

Generally, standard rated outputs of motors are based on temperature rise (75°C) for insulation class B, where the temperature rise is more, old motors should be replaced with higher class °F (95°C).

16.2.3   Misalignment and vibrations

The principal causes for bearing failure could be misalignment of motor with driven equipment and subsequent vibrations, cumulative errors in tolerances, wrong size or type of bearing, ingress of foreign matter, inadequate and contaminated lubrication, etc. It accounts for 20–25% of failures. The abnormal bearing noise because of misalignment and/or vibration, serves as an early warning for further trouble, e.g., bearing seizure. Hence, alignment should be checked and confirmed by a senior person before starting the equipment, which had been taken out for maintenance.

16.2.4   Corrosive fluids/water entry

High corrosive atmosphere in a chemical plant causes rapid deterioration leading to early failure. This accounts for 12–15% of breakdowns of motors. Leakages of chemicals/corrosive materials from the driven equipment are not only safety hazards for plant operations but potential problems for the motors as well. Hence, all such leakages from driven equipments and pipelines should be arrested immediately. To avoid such failures, motors could be provided with expeller on the shaft to prevent entry of any fluid in to the motor.

16.2.5   Higher temperature/overloading

15–20% of failures of LT induction motors take place because of higher temperature, overloading, electrical load inbalance, incorrect voltage and frequency, motor winding with loose connections, etc. Motors are intended for operation in maximum ambient temperature of 45°C. The motors near dryers, furnaces driving hot air blowers, fans, etc., in foundry and textile plants, which are operated at higher temperatures, should be derated as below to avoid premature failure. Figure 16.1 shows the location of motor failing frequently due to higher temperature because of steam leakage from the inspection door.

Environment temp °C →	45	50	55	60	65	75	85
Permitted output → as % of rated output	100	96.5	93.0	90.0	86.5	79.0	70.0

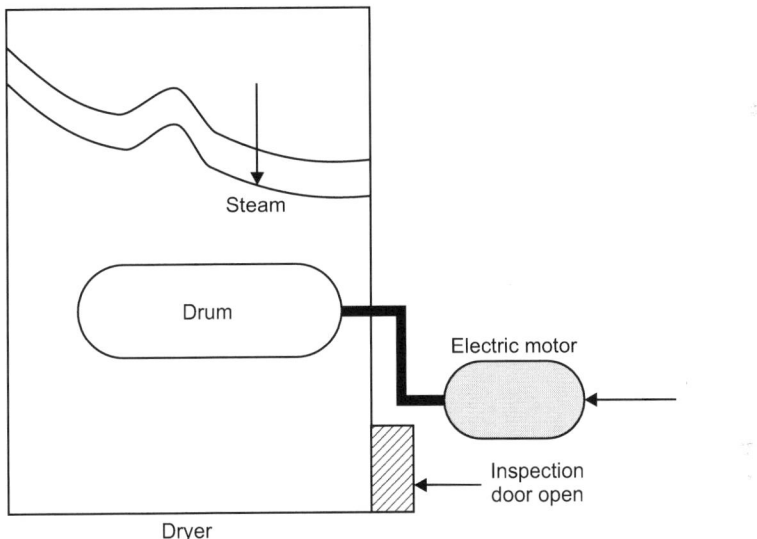

Figure 16.1: Showing wrong location of the motor.

Overloading and jamming of motors is mainly due to change in the process parameters. Frequent load variations for more than 15 seconds, due to process upsets or jamming because of mechanical problems, damage the insulation which results in motor break down. Hence, overloading due to change in process parameters should be identified and rectified. Many faults develop because of frequent motor starts. It causes the motor to carry starting loads and currents for a long time leading to overheating. In such cases, direct on line start current can be reduced with soft start circuits. Electronic timed over load protection can be also be used.

16.2.6   Inadequate protection

In a three phase motor blowing of a fuse, by burning out or mechanical interruption of one of the supply leads, by a faulty switch contact or circuit breaker could cause single phasing. When the motor runs on single phase, it gradually shows signs of distress, increase in noise, decrease in speed and overheating occurs. If adequate protection is not provided, the motors would burn out. The burn out is caused by an abnormal increase in the current flowing in the motor winding.

16.2.7   Operating parameter

Monitoring the operating parameters of electrical motors is essential to prevent forced plant outages due to any of the above reasons. This would give a timely and reliable evaluation of the motor condition. It could help plan maintenance and prevent failure, thereby avoid expensive repairs.

The tests and inspections could be carried out at regular intervals during plant overhauls. The motor operating conditions could be recorded and analysed for failure of each motor and rectified to avoid re-replacing them by energy efficient ones can save energy. Energy efficient motors are manufactured as per IEEMA 19–2000 and cost 20–25% more. Their cost can be recovered with 1500–2000 running hours, i.e., less than a year.

Most of the motors in the plants are running around 60–70% loads, which results in higher losses, whereas energy efficient motors have constant efficiency between 50–100% of full load. This means savings in energy at lower loads also. Hence actual load at operating parameters should be checked with efficiency curves given by the suppliers.

16.3   Energy-efficiency improvement opportunities in electric motors

When considering energy-efficiency improvements to a facility's motor systems, a systems approach incorporating pumps, compressors and fans must be used in order to attain optimal savings and performance.

In the following, considerations with respect to energy use and energy saving opportunities for a motor system are presented. Pumping, fan and compressed air systems are discussed in addition to the electric motors.

16.3.1   Motor management plan

A motor management plan is an essential part of a plant's energy management strategy. Having a motor management plan in place can help companies realise long-term motor system energy savings and will ensure that motor failures are handled in a quick and cost effective manner.

The following are the key elements for a sound motor management plan:

1. Creation of a motor survey and tracking programme.
2. Development of guidelines for proactive repair/replace decisions.
3. Preparation for motor failure by creating a spares inventory.
4. Development of a purchasing specification.
5. Development of a repair specification.
6. Development and implementation of a predictive and preventive maintenance programme.

Maintenance

The purposes of motor maintenance are to prolong motor life and to foresee a motor failure. Motor maintenance measures can therefore be categorised as either preventative or predictive. Preventative measures, include voltage imbalance minimisation, load consideration, motor alignment, lubrication and motor ventilation.

Some of these measures are further discussed below.

Note that some of them aim to prevent increased motor temperature which leads to increased winding resistance, shortened motor life and increased energy consumption. The purpose of predictive motor maintenance is to observe ongoing motor temperature, vibration and other operating data to identify when it becomes necessary to overhaul or replace a motor before failure occurs. The savings associated with an ongoing motor maintenance programme could range from 2% to 30% of total motor system energy use.

16.3.2   Energy-efficient motors

Energy-efficient motors reduce energy losses through improved design, better materials, tighter tolerances and improved manufacturing techniques. With proper installation, energy-efficient motors can also stay cooler, may help reduce facility heating loads and have higher service factors, longer bearing life, longer insulation life and less vibration. The choice of installing a premium efficiency motor strongly depends on motor operating conditions and the life cycle costs associated with the investment. In general, premium efficiency motors are most economically attractive when replacing motors with annual operation exceeding 2000 hr/year. Sometimes, even replacing an operating motor with a premium efficiency model may have a low payback period.

Rewinding of motors

In some cases, it may be cost-effective to rewind an existing energy efficient motor, instead of purchasing a new motor. As a rule of thumb, when rewinding costs exceed 60% of the costs of a new motor, purchasing the new motor may be a better choice. When repairing or rewinding a motor, it is important to choose a motor service center that follows best practice motor rewinding standards in order to minimise potential efficiency losses. When best rewinding practices are implemented, efficiency losses are typically less than 1%. Software tools such as Motor Master can help identify attractive applications of premium efficiency motors based on the specific conditions at a given plant.

Proper motor sizing

It is a persistent myth that oversized motors, especially motors operating below 50% of rated load, are not efficient and should be immediately replaced with appropriately sized energy efficient units. In actuality, several pieces of information are required to complete an accurate assessment of energy savings. They are the load on the motor, the operating efficiency of the motor at that load point, the full-load speed [in revolutions per minute (rpm)] of the motor to be replaced and the full-load speed of the downsized replacement motor.

The efficiency of both standard and energy efficient motors typically peaks near 75% of full load and is relatively flat down to the 50% load point. Motors in the larger size ranges can operate with reasonably high efficiency at loads down to 25% of rated load. There are two additional trends: larger motors exhibit both higher full- and partial-load efficiency values and the efficiency decline below the 50% load point occurs more rapidly for the smaller size motors. Software packages such as Motor Master can aid in proper motor selection.

Adjustable speed drives (ASDs)

Adjustable-speed drives better match speed to load requirements for motor operations and therefore ensure that motor energy use is optimised to a given application. As the energy use of motors is approximately proportional to the cube of the flow rate, relatively small reductions in flow, which are proportional to pump speed, already yield significant energy savings.

Power factor correction

Power factor is the ratio of working power to apparent power. It measures how effectively electrical power is being used. A high power factor signals efficient utilisation of electrical power, while a low power factor indicates poor utilisation of electrical power. Inductive loads like transformers, electric motors and HID lighting may cause a low power factor. The power factor can be corrected by minimising idling of electric motors (a motor that is turned off consumes no energy), replacing motors with premium-efficient motors and installing capacitors in the AC circuit to reduce the magnitude of reactive power in the system.

Minimising voltage unbalances

A voltage unbalance degrades the performance and shortens the life of three-phase motors. A voltage unbalance causes a current unbalance, which will result in torque pulsations, increased vibration and mechanical stress, increased losses and motor overheating, which can reduce the life of a motor's winding insulation. Voltage unbalances may be caused by faulty operation of power factor correction equipment, an unbalanced transformer bank, or an open circuit. A rule of thumb is that the voltage unbalance at the motor terminals should not exceed 1% although even a 1% unbalance will reduce motor efficiency at part load operation. A 2.5% unbalance will reduce motor efficiency at full load operation. By regularly monitoring the voltages at the motor terminal and through regular thermographic inspections of motors, voltage unbalances may be identified. It is also recommended to verify that single-phase loads are uniformly distributed and to install ground fault indicators as required. Another indicator for voltage unbalance is a 120 Hz vibration, which should prompt an immediate check of voltage balance. The typical payback period for voltage controller installation on lightly loaded motors is 3 years.

16.3.3 Energy savings

Energy savings can be achieved by either using energy efficient motors or reducing motor failures and losses, which is possible by:

1. Running the motors at rated load.
2. Reducing copper losses by increasing the volume of copper wire.
3. Reducing iron losses by using superior grade of stampings and/or longer core length.

Failure of insulation decreases the properties of materials because of heating. This decreases the efficiency by about 0.5%, every time re-insulation is done. Observations have shown that the contractors do re-insulation of motors with the materials available with them. No efficiency tests are carried out in their shop after the repairs. Hence the motors, which are repaired 5–6 times, have more losses and lower efficiency. Therefore, monitoring of motor failures and regular checks and upkeep of motors as recommended by the manufacturers and applicable standards. Regular monitoring of the working conditions and general inspection can discover faults like leakage of lubricants burning smell, higher temperature, noise and looseness of bolts, etc., which with timely rectification could avoid motor failures in future and save energy.

16.4   Gear efficiency

Efficiency of a speed reducer is an important selection factor that is often overlooked. It shouldn't be. In many cases, high-efficiency gearing cuts the cost of drives and their operation. Because they are widely used with industrial equipment, speed reducers and gear motors can significantly impact your drive costs.

16.4.1 Reducer type determines efficiency

Though reducer efficiency may vary slightly from one manufacturer to another, the way in which the gears intersect and mesh mainly determines speed reducer efficiency. This efficiency ranges from 49 to 98%, depending on the type of reducer and number of reduction stages it contains. Here's a brief description of some common types and their relative efficiencies.

Worm gear: In these widely used speed reducers, a worm gear drives a worm wheel to provide output motion at a right angle to the motor shaft (Fig. 16.2). The worm gear and worm wheel have non-intersecting, perpendicular axes and the meshing action between gears occurs over a relatively large contact area. This meshing action consists primarily of a sliding motion that creates friction between the gears.

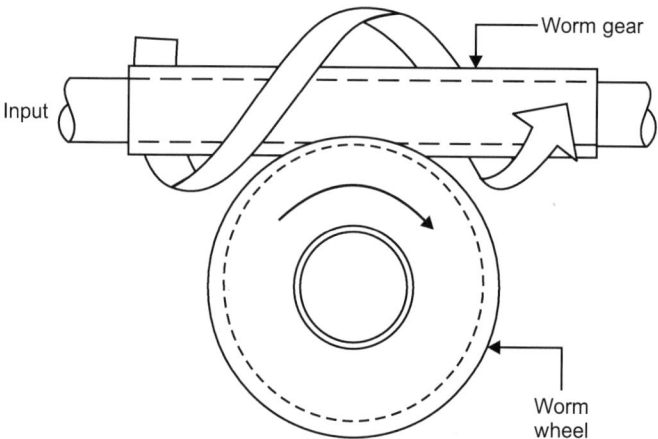

Figure 16.2: Worm-gear set for a speed reducer.

Efficiency of a worm-gear speed reducer depends (in part) on its speed-reduction ratio. High-ratio units have a smaller gear-tooth lead (helix) angle, which causes more surface contact between them. This higher contact causes higher friction and lower efficiency. Typical worm-gear efficiencies range from 49% for a 300:1, double-reduction ratio, up to 90% for a 5:1, single-reduction ratio. For this reason, these units are usually more suitable for low ratios.

Helical worm: The arrangement of gear stages in a speed reducer also affects its efficiency. In helical-worm speed reducers, a set of helical gears connected to the motor shaft drives a worm-gear set (Fig. 16.3). The helical-gear set

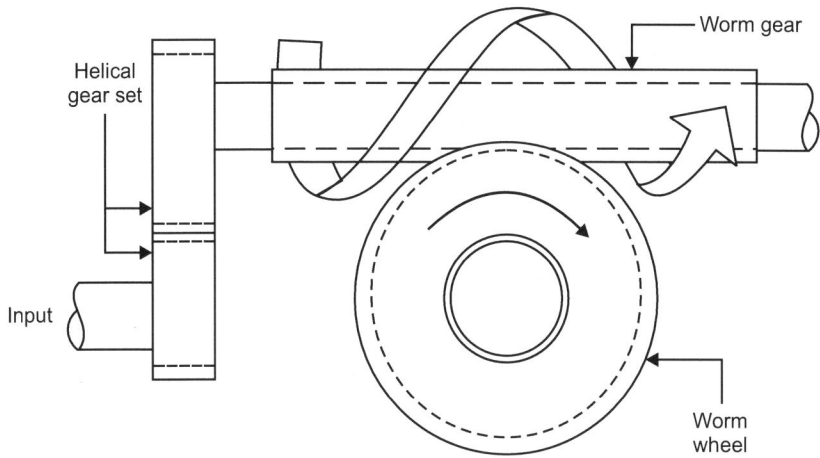

Figure 16.3: Helical worm-gear set.

reduces input speed (from the motor) to the wormgear set. This keeps the worm-gear reduction ratio (and size) low to maximise its efficiency.

The combined efficiency of helical and worm gears ranges from 79% for a 300:1 speed-reduction ratio to 90% for a 5:1 ratio.

Helical bevel: As with worm and helical- worm units, the output shaft of this reducer type is at a right angle to the motor shaft (Fig. 16.4). And like helical-worm speed reducers, a set of helical gears usually makes the first speed reduction. Here, spiral-bevel gears are mounted with intersecting axes, an arrangement that minimises friction between the gears to provide 94 to 97% efficiency.

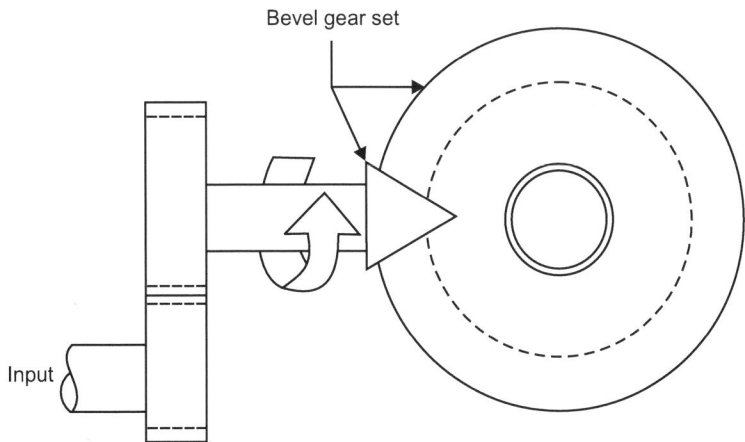

Figure 16.4: Helical bevel gears.

The primary drawback to helical-bevel speed reducers is their higher cost than worm-gear units, especially for small ratios such as 5:1 or 10:1. But the added cost is sometimes offset in high-reduction- ratio units because their high efficiency allows the use of either a smaller reducer, motor, or both.

In-line helical: These speed reducers are often the best value for applications in which the motor and reducer are coupled in-line with the driven shaft (Fig. 16.5). Their efficiency usually ranges between 95 and 98%, regardless of speed ratio. Cost is competitive with worm-gear speed reducers where the center distance is 2-in. or more.

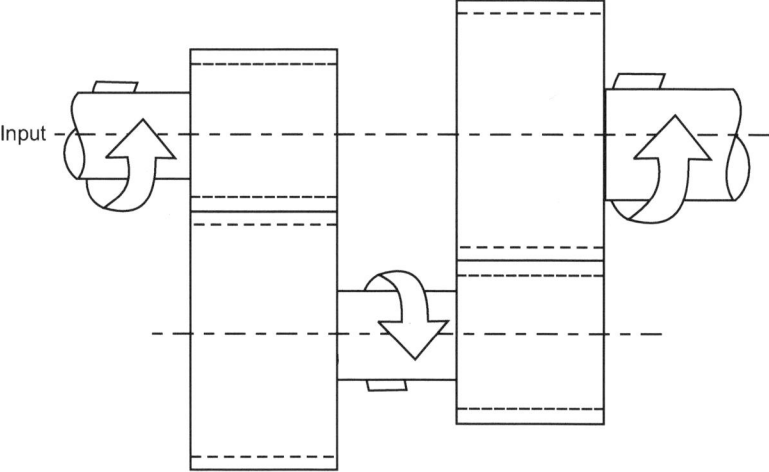

Figure 16.5: In-line helical gears have parallel axes where input and output shafts are on the same centerline.

Helical-gear sets have parallel axes between gear teeth and small contact areas, which keeps power-robbing friction and heat low.

16.4.2 Calculating efficiency

To compare different types of speed reducers, first calculate the required load for your application and the efficiencies of the reducers being considered.

Load requirement: The key to proper speed-reducer selection is to determine the actual torque or power required to drive the load. First measure the torque (lb-in.) or power (hp) needed to drive the load at its input shaft.

Next, multiply the torque or power value by appropriate service factors to compensate for unusual operating conditions such as shock loads, frequent stopping and starting and high temperature. These service factors are normally listed in manufacturers' catalogs. The American Gear Manufacturers Association (AGMA) also publishes service factors for different classes of service in their standards. For example, if a driven equipment shaft requires 2 hp and it's subject to moderate shock loads, apply a 1.4 service factor. If the ambient temperature is between 100 and 115°F, use a 1.15 service factor. Thus:

$$P_R = P_0 \times SF_S \times SF_T \qquad \ldots(16.1)$$

$$= 2 \times 1.4 \times 1.15 = 3.22 \text{ hp}$$

Once you've applied the recommended service factors to compensate for loading and temperature, compare the efficiency of the speed-reducers being considered based on the actual load requirement.

Efficiency: Unfortunately, many speed-reducer manufacturers do not publish efficiency ratings. If their catalog lists the input power, torque, or wattage as well as comparable output values, you can calculate efficiency by:

$$E = \frac{P_1}{P_0} \qquad \ldots(16.2)$$

If you need to convert power, torque, or wattage values to make a comparison, use these formulas:

$$P_m = \frac{T \times S}{63025} \qquad \ldots(16.3)$$

$$P_e = P_m \times 0.75 \qquad \ldots(16.4)$$

Input power requirement in gears

Once the service-factored power and the gear efficiency are known, determine the required motor power by:

$$P_M = P_0 \times \frac{1}{E_{SR}} \qquad \ldots(16.5)$$

$$= 2 \times \frac{1}{0.95} = 2.1 \text{ hp}$$

As a rule of thumb, choose the next larger motor, in an example, a 3-hp motor.

Energy consumption

Nomenclature:

P_g	=	Power required, hp	P_m	=	Power (mechanical), hp
P_o	=	Output power, hp	T	=	Torque, lb-in
		(or lb-in, or kW)	S	=	Speed, rpm
SF_S	=	Service factor for shock load	P_e	=	Power (electrical), kW
SF_T	=	Service factor for high temperature	P_M	=	Motor power, hp
E	=	Efficiency, %	E_{SR}	=	Speed reducer efficiency, %
Pt	=	Input power, hp	P_C	=	Power cost, $

(or lb-in. or kW) 　　　　　　　 H = Time, hr

　　　　　　　　　　　　　　　　C_K = Cost, \$/kW-h

To estimate the cost of power consumed by a reducer, convert the motor power requirement to kW. Then multiply this value by the annual number of operating hours and the local cost per kW-h:

$$P_C = P_M \times 0.75 \times H \times C_K \qquad \qquad ...(16.6)$$

$$= 2.1 \times 0.75 \times 4000 \times 0.08$$

$$= \$504$$

Because this calculation doesn't consider motor efficiency and power factor, use it only to compare the energy consumption of different speed reducers, not the motor-reducer combination.

Costs

Gear efficiency affects both drive component (speed reducer and motor) costs and operating costs, as demonstrated by examining two common speed-reducer types: a concentric or in-line helical-gear unit and a right angle worm-gear unit. Both units provide a 60:1 ratio and are rated with a 1.0 service factor for uniform loads and normal ambient temperature. They are driven by a 1750-rpm, 230/460- V, 3-phase TEFC motor.

As Table 16.1 shows, the higher efficiency of the in-line helical reducer enables using a ½-hp smaller motor.

Table 16.1: Motor size in proportion to helical and worm gear.

	In-line helical gear	*Worm gear*
Load required, hp	1	1
Efficiency, %	96	62
Input power required, hp	1.04	1.61
Recommended motor power, hp	1.5	2.0

Though gear efficiency can cut costs through the use of a smaller reducer and motor, it also minimises energy costs.

Not every application produces the same results. For example, worm-gear speed reducers with ratios up to 30:1, or those with 2-in. center distances or less, are often a better value than helical reducers. Such worm-gear units offer about 78% efficiency, compared to 62%, because the worm gear and wheel are smaller and generate lower friction losses.

Environmental benefit

Efficient speed reducers can achieve significant savings, both in drive component costs and energy costs. But there is another good reason to use them: to reduce health risks from greenhouse gases in our environment. According to the U.S. department of energy, industrial electric motor systems account for 1/12 of all greenhouse gas emissions from fossil fuel power plants. Therefore, cutting motor power consumption with efficient speed reducers will help to ease this global problem.

16.4.3   Gear lubricants

The increase of gear efficiency harbours a frequently overlooked potential for increasing the efficiency of a machine as a whole. A very direct and effective way of increasing power transmission efficiency—which goes along with excellent wear protection—is a change over from mineral-oil-based to synthetic lubricants. Synthetic lubricants based on polyalphaolefin, ester or polyglycol oils, for example, have proven to reduce energy costs and in addition extend the service life of gears. The possible extent of efficiency increase, however, depends on the type of gears: while gears featuring a low percentage of sliding friction, such as spur or bevel gears, offer a relatively low potential, gears with a high percentage of sliding friction enable considerable improvements.

A particularly positive effect can be noted when worm gears are switched over to polyglycol oil: their efficiency has been increased by up to 35%. In addition, their lifetime can be extended tenfold. A conversion to synthetic oils offers an enormous potential for savings especially in facilities where a large number of gears are operated—for example in logistics centers, filling stations, breweries or airports. The example in the sidebar illustrates how several million euros can be saved at a large airport.

Tribological factors are decisive in attaining the maximum performance of a machine and its components. When choosing a lubricant for a gearbox or machine, therefore, design engineers should be aware of the characteristics of the various types of lubricants and know how to use them. While, as a rule, synthetic special lubricants tend to be more expensive than mineral oils in terms of the sales price, they pay off after a short time when taking into account efficiency, oil change intervals, oil consumption and the longer lifetime of lubricated components. With such lubricants, gear manufacturers offer their customers the added benefit of lower operating costs.

Minimising wear

Polyglycol oils offer the best wear protection to worm gears. They can help to extend the lifetime of a worm gearbox significantly compared with a mineral oil. Consequently, the worm wheel survives longer with the same load or the output torque can be increased without dimensional changes. Additional benefits for machine operators are cost savings due to longer maintenance intervals, a lower risk of equipment failure and minimised downtime.

Maximum energy efficiency in gear units

Maximum energy efficiency of a gear unit means that it produces the highest possible output power for a given input power. The energy lost in the process manifests itself in the form of heat, for example in bearings, O-ring seals or gear wheels. As gear efficiency increases, its temperature will go down. This has a number of positive effects: a decreasing temperature not only extends the oil life, but the service life of seals as well. This in turn reduces the risk of leakage. Another benefit is that fans or air conditioners in production facilities might be switched off, which is another contributor to lower energy costs and a better CO_2 balance.

16.5 Fans and blowers

Fans and blowers provide air for ventilation and industrial process requirements. Fans generate a pressure to move air (or gases) against a resistance caused by ducts, dampers, or other components in a fan system. The fan rotor receives energy from a rotating shaft and transmits it to the air. Fans fall into two general categories: centrifugal flow and axial flow. In centrifugal flow, airflow changes direction twice - once when entering and second when leaving (forward curved, backward curved or inclined, radial). In axial flow, air enters and leaves the fan with no change in direction (propeller, tubeaxial, vaneaxial).

Blower is a rotary, positive displacement type of machine used to move gas and air and is used in a variety of methods. Rotary blowers and positive displacement blowers are just two of several types that are available. Because fan systems often directly support production processes, many fans operate continuously. These long run times translate into significant energy consumption and substantial annual operating costs. The operating costs of large fans are often high enough that improving fan system efficiency can offer a quick payback.

1. Inspite of this, facility personnel often do not know the annual operating costs of an industrial fan, or how much money could be saved by improving fan system performance.

2. Operating costs of fan systems primarily include electricity and maintenance costs. Of these two components, electricity costs can be determined with simple measurements. In contrast, maintenance costs are highly dependent on service conditions and need to be evaluated case-by-case. A particularly useful method of estimating these costs is to review the maintenance histories of similar equipment in similar applications.

3. The cost of operating a fan system is affected by the amount of time and the percentage of full capacity (load factor) at which the fan motor operates. Because the fan system does not usually operate at rated full load all the time, an estimate of its average load factor must be made.

16.5.1   Determination of the load factor

The load factor of a fan system can be determined by listing the number of operating hours at each level of output over a typical plant cycle like one week. By multiplying the number of hours by the level of output, adding the results and dividing by the total number of hours in the entire period, one obtains the average load factor of the fan system.

16.5.2   Calculating electricity consumption

Electricity consumption can be determined by several methods, including:

1. Using motor nameplate data.
2. Using direct electrical measurements.
3. Using fan performance curve data.

In systems with widely varying operating conditions, simply taking data once will probably not provide a true indication of fan energy consumption. It is better to use data for several operating points and use a weighted average based on hours of operation at each point.

The motors used on most fans have a 1.15 continuous service factor. This means that a motor with a nominal nameplate rating of 150 brake horse power (bhp) may be operated continuously up to 172.5 bhp, although motor efficiency drops slightly above the rated load. Using nameplate data to calculate energy costs on motors that operate above their rated loads will understate actual costs. A more accurate way to determine electricity consumption requires taking electrical measurements of motor current, voltage and power factor over a range of operating conditions.

16.5.3   Matching fans/blowers to motor

Fans are usually selected to match the maximum pressure and flow requirement of the system. Although the system may not be required to operate at the maximum conditions all the time, the fan-motor combination must be capable of delivering the required air flow when needed.

1. The motor selected for the drive system must be capable of supplying the fan with the required driving power.
2. Fans are typically driven by alternating current (AC) motors. In industrial fan applications, the most common motor type is the squirrel-cage induction motor, selected for its durability, low cost, reliability and low maintenance requirements. These motors are commonly available with 2 or 4 poles which, on a 60- hertz system, translate to nominal operating speeds of 3600 revolutions per minute (rpm) and 1800 rpm, respectively.
3. Although motors with 6 poles or more are used in slower fan systems, they are relatively expensive. Motors can operate safely over long periods

of time above the rated output or with a 'service factor' above 1. Service factors range from 1.1 to 1.15, meaning that the motors can safely operate at loads between 110 and 115% of their output power rating.

16.5.4 Energy saving opportunities

Minimising demand on the fan

1. Minimising excess air level in combustion systems to reduce FD fan and ID fan load.
2. Minimising air in-leaks in hot flue gas path to reduce ID fan load, especially in case of kilns, boiler plants, furnaces, etc. Cold air in-leaks increase ID fan load tremendously, due to density increase of flue gases and in-fact choke up the capacity of fan, resulting as a bottleneck for boiler/furnace itself.
3. In-leaks/out-leaks in air conditioning systems also have a major impact on energy efficiency and fan power consumption and need to be minimised.

The findings of performance assessment trials will automatically indicate potential areas for improvement, which could be one or a more of the following:

1. Change of impeller by a high efficiency impeller along with cone.
2. Change of fan assembly as a whole by a higher efficiency fan.
3. Impeller derating (by a smaller dia impeller).
4. Change of metallic/glass reinforced plastic (GRP) impeller by the more energy efficient hollow FRP impeller with aerofoil design, in case of axial flow fans, where significant savings have been reported.
5. Fan speed reduction by pulley dia modifications for derating.
6. Option of two speed motors or variable speed drives for variable duty conditions.
7. Option of energy efficient flat belts, or, cogged raw edged V belts, in place of conventional V belt systems, for reducing transmission losses.
8. Adopting inlet guide vanes in place of discharge damper control.
9. Minimising system resistance and pressure drops by improvements in duct system.

16.5.5 Energy-efficiency in compressed air systems

Air compressor is a compressor that takes in air at atmospheric pressure and delivers it at a higher pressure. More than 85% of the electrical energy input to an air compressor is lost as waste heat, leaving less than 15% of the electrical energy consumed to be converted to pneumatic compressed air energy. This makes compressed air an expensive energy carrier compared to other energy carriers. Many opportunities exist to reduce energy use of compressed air systems. For optimal savings and performance, it is recommended that a systems approach is used. In the following, energy saving opportunities for compressed air systems are presented. Also, Energy Assessment for Compressed Air Systems (ASME) has published a standard that covers the assessment of compressed air systems that are defined as a group of subsystems of integrated sets of components for consistent, reliable and efficient use of energy. In this standard the procedure of conducting a detailed energy assessment of the compressed air system as well as the energy efficiency opportunities are described.

1. Reduction of demand.
2. Maintenance.
3. Monitoring.
4. Reduction of leaks (in pipes and equipment).
5. Electronic condensate drain traps (ECDTs).
6. Reduction of the inlet air temperature.
7. Maximising allowable pressure dew point at air intake.
8. Optimising the compressor to match its load.
9. Proper pipe sizing.
10. Heat recovery.
11. Installing adjustable speed drives (ASDs).

Reduction of demand: Because of the relatively expensive operating costs of compressed air systems, the minimum quantity of compressed air should be used for the shortest possible time, constantly monitored and reweighed against alternatives.

Maintenance: Inadequate maintenance can lower compression efficiency, increase air leakage or pressure variability and lead to increased operating temperatures, poor moisture control and excessive contamination. Better maintenance will reduce these problems and save energy.

Monitoring: Maintenance can be supported by monitoring using proper instrumentation, including following:

1. Pressure gauges on each receiver or main branch line and differential gauges across dryers, filters, etc.
2. Temperature gauges across the compressor and its cooling system to detect fouling and blockages.
3. Flow meters to measure the quantity of air used.
4. Dew point temperature gauges to monitor the effectiveness of air dryers.
5. kWh meters and hours run meters on the compressor drive.

Reduction of leaks (in pipes and equipment): Leaks cause an increase in compressor energy and maintenance costs.

The most common areas for leaks are:

1. Couplings.
2. Hoses.
3. Tubes.
4. Fittings.
5. Pressure regulators.
6. Open condensate traps and shut-off valves.
7. Pipe joints.
8. Disconnects.
9. Thread sealants.

Quick connect fittings always leak and should be avoided: In addition to increased energy consumption, leaks can make pneumatic systems/equipment less efficient and adversely affect production, shorten the life of equipment, lead to additional maintenance requirements and increased unscheduled downtime. A typical plant that has not been well-maintained could have a leak rate between 20 and 50% of total compressed air production capacity. Leak repair and maintenance can sometimes reduce this number to less than 10%. Overall, a 20% reduction of annual energy consumption in compressed air systems is projected for fixing leaks. A simple way to detect large leaks is to apply soapy water to suspect areas. The best way is to use an ultrasonic acoustic detector, which can recognise the high frequency hissing sounds associated with air leaks.

Electronic condensate drain traps (ECDTs): Due to the necessity to remove condensate from the system, continuous bleeding, achieved by forcing a receiver drain valve to open, often becomes the normal operating practice, but is extremely wasteful and costly in terms of air leakage. ECDTs offer improved reliability and are very efficient as virtually no air is wasted when the condensate is rejected. The payback period depends on the amount of leakage reduced and is determined by the pressure, operating hours, the physical size of the leak and electricity costs.

Reduction of the inlet air temperature: Reducing the inlet air temperature reduces energy used by the compressor. In many plants, it is possible to reduce this inlet air temperature by taking suction from outside the building. Importing fresh air has paybacks of up to 5 years, depending on the location of the compressor air inlet. As a rule of thumb, each 3°C reduction will save 1% compressor energy use.

Maximising allowable pressure dew point at air intake: Choose the dryer that has the maximum allowable pressure dew point and best efficiency. A rule of thumb is that desiccant dryers consume 7–14% and refrigerated dryers consume 1–2% of the total energy of the compressor.

Consider using a dryer with a floating dew point. Note that where pneumatic lines are exposed to freezing conditions, refrigerated dryers are not an option.

Optimising the compressor to match its load: Plant personnel have a tendency to purchase larger equipment than needed driven by safety margins or anticipated additional future capacity. Given the fact that compressors consume more energy during part-load operation, this is something that should be avoided. Some plants have installed modular systems with several smaller compressors to match compressed air needs in a modular way. In some cases, the pressure required is so low that the need can be met by a blower instead of a compressor which allows considerable energy savings, since a blower requires only a small fraction of the power needed by a compressor.

Proper pipe sizing: Pipes must be sized correctly for optimal performance or resized to fit the compressor system. Inadequate pipe sizing can cause pressure losses, increase leaks and increase generating costs. Increasing pipe diameter typically reduces annual compressor energy consumption by 3%.

Heat recovery: As already mentioned, more than 85% of the electrical energy used by an industrial air compressor is converted into heat. A 150 hp compressor can reject as much heat as a 90 kW electric resistance heater or a 422 MJ/hour natural gas heater when operating. In many cases, a heat recovery unit can recover 50–90% of the available thermal energy for space heating, industrial process heating, water heating, makeup air heating, boiler makeup water preheating, industrial drying, industrial cleaning processes, heat pumps, laundries or preheating aspirated air for oil burners. With large water-cooled compressors, recovery efficiencies of 50–60% are typical. When used for space heating, the recovered heat amount to 20% of the energy used in compressed air systems annually. 'Paybacks are typically less than one year'.

In some cases, compressed air is cooled considerably below its dew point in refrigerated dryers to condense and remove the water vapour in the air. The waste heat from these after coolers can be regenerated and used for space heating, feed water heating or process-related heating.

Installing adjustable speed drives (ASDs): When there are strong variations in load and/or ambient temperatures there will be large swings in compressor load and efficiency. In those cases, installing an ASD may result in attractive payback periods. 'Implementing adjustable speed drives in rotary compressor systems has saved 15% of the annual compressed air system energy consumption'.

Energy efficient pumps and V-belts

17.1   Introduction

Energy savings of pumps highly depend on correct pump selection and sizing. The world's energy balance is one of the biggest challenges of our era. On the one hand, there are huge efforts to generate 'green energy' from renewable sources. But it still needs to be considered that the most ecological energy is the energy not needed. So on the other hand are a number of measures to reduce energy consumption.

Continuously rising energy costs have increased the demands of pump manufacturers to provide energy efficient solutions for fluid transfer. Besides the hydraulic optimisation of the units, a number of new developments in the field of high efficient electrical drives and powerful control systems have been launched in order to meet those requirements.

We often underestimate the savings, which can be already generated with the dimensioning of the pipeline and the right duty point specification of the pump within the planning phase.

World energy consumption of pumps: Pumping systems account for nearly 20% of the world's electrical energy demand and range from 25–50% of the energy usage in certain industrial plant operations.

Considering the major role of the power usage of pumps, it is necessary to reduce the world's energy consumption. To use this challenge in the best way, different options of energy saving applications have to be analysed.

17.2   Methodology adopted for performance evaluation of pumping system

The prime objective of pump performance evaluation is to estimate the actual efficiency of the pump. The efficiency of the pump is directly proportional to the flow and head. Pump performance analysis can be carried out by knowing the liquid flow rate, liquid pressure, motor input power and pump rotating speed. Table 17.1 gives brief information about different operating parameters and typical measuring instruments required in pumping sets.

By knowing the liquid flow, operating head and motor input power, pump efficiency can be evaluated by the following expression:

$$\eta_{pump} = (Q \times H \times \rho)/(102 \times kW \times \eta_{motor})$$

where,

η_{pump}	=	Operating efficiency of pump, in per cent
Q	=	Capacity, in m³/s
H	=	Head, in m
P	=	Density of liquid handled, in kg/m
kW	=	Power consumed by motor, in kW
η_{motor}	=	Efficiency of motor, in per cent

Table 17.1: Brief information about different operating parameters and typical measuring instruments required for pumping sets.

Parameters	*Typical measurement instruments*
Liquid flow rate: Comparing liquid rates in different parts of the system can help pinpoint leaks, high pipe friction and real time pumping requirements.	Differential pressure devices, such as orifice meters and venturi meters.
	Velocity flow meters, such as pitot tubes
	Open flow meters
	Positive displacement meters
	Ultrasonic flow meters
Liquid pressure: Monitoring water pressure can help find leaks, reduce unnecessary pumping, and maintain constant service.	Bourdon tubes
	Bellows
	Diaphragm gauges
	Piezo-resistive transducers
Motor input power: Input power readings can help determine if a motor is operating at its optimal loading (efficiency).	Ammeters
	Voltmeter
	Power factor meters
Pump rotating speed: Rotating speed data can help determine if a motor is operating at its optimal loading (efficiency)	Stroboscope

17.2.1 Possible energy conservation options in pumping systems

1. Selection of low/high head pumps.
2. Capacity regulation by opting for variable speed drives (VSD) in place of throttling control.
3. Avoiding recirculation/by-pass by opting for two way valves and VSD.
4. Adequate sizing of suction and discharge pipe.
5. Parallel operation of pumps.
6. Trimming of impellers.
7. Reduction of frictional losses by selection of low friction pipes.

8. Reduction of frictional losses by adequate sizing of pipes.
9. Downsizing of pumps with low capacity pumps.
10. Providing sufficient net positive suction head (NPSH).
11. Installation of overhead tanks.
12. Matching of correct sized motor for the pumps.

Selection of low/high head pumps: In process industries, pumping systems are designed to deliver through the shortest possible route by minimising the number of bends, valves, etc. However, when new pumps are required for the system, an adhoc approach is taken and pumps with a higher head (by about 20%) are installed. This leads to operation of pumps away from the best efficiency point. Being continuous duty pumps, when they are operated inefficiently, huge amount of energy is lost considering the life cycle of the pumps. The buyer or plant personnel should always install a pump as close to the system head requirements as possible to avoid inefficient operations. Same is the case for high head applications too.

Sometimes when system head requirements are higher, plant people install a low head pump for given duty. This results in low output of the pump, over-loading of the motor and operation of pumps at lower efficiency leading to high power consumption. Careful analysis of the system requirements have to be assessed while installing new pumps.

Capacity regulation by opting for variable speed drives (VSD) in place of throttling control: Quite often, there is variable flow requirement from pumps due to changes or flexibilities in output, varying water requirements, etc. In such cases normally plant personnel throttle the discharge valve of the pumps to meet the water requirement. Under such circumstances, a more accurate and energy saving method is to install variable speed drive for the pumps during variable requirements, the speed of the pump can be varied as per the requirements. As per affinity laws, speed is proportional to the cube of power, hence any reduction in speed will result in reduction in power consumption by one third thereby resulting in power savings.

Avoiding recirculation/by-pass by opting for two way valves and VSD: In pump systems sometimes a bypass valve is provided, which returns a portion of the water back to the tank and the balance of the water required is only utilised for process requirements. This method of flow control is inefficient as some portion of water is recirculated consuming energy. Providing two way valves and installing variable speed drives will result in pumps operating specific to process requirements, thereby resulting in energy savings.

Adequate sizing of suction and discharge pipe: When pump manufacturers design a pump the inlet and the outlet pipe dimensions connected to the casing are specified. During installation of the pump it so happens that the guidelines are not strictly followed which results in pressure losses at the inlet and outlet of the pump in excess of what would have been if the pipes were designed adequately. Hence it is advisable to size the suction and discharge pipes as specified by the manufacturer. This will help in operating the pump at its peak efficiency.

Parallel operation of pumps: Pumps are operated in parallel if different requirements exist throughout the day. This is so when capacity of pumps is large and also when variable speed drives don't suit the system requirements due to high static head. In such cases parallel operation of pumps is the solution. However, in most of the industries, parallel operation of pumps is carried out to meet the water requirements constantly round the clock throughout the year. Any parallel operation of pumps with three or four pumps in combination leads to efficiency loss by 15% to 20%. Installing higher capacity pumps will improve the operating efficiency and hence reduction in power consumption is achievable for the same application.

Trimming of impellers: Trimming of impellers is desired if already a higher capacity pump is installed and flow required is marginally less compared to the design parameters. In such cases trimming of impellers will result in required output which also results in energy savings. Here care has to be taken that trimming of impellers should be done upto 5% to 7% of the original diameter of the impeller. Trimming in excess of 7% will result in recirculation of water inside the casing thereby resulting in low output. In case the output of the pump required is 20% to 30% less than the pump installed, it is advisable to downsize the pump to required capacity. Another concern while selecting a new pump is that at times the required capacity of the pump impeller will not be available with the manufacturer. In such cases, the manufacturer trims the original impeller to suit the required capacity as specified by the buyer. This leads to inefficiency at the initial stage itself.

This issue has to be thoroughly discussed with the manufacturer while selecting new pumps. As far as possible it is better to opt for a correct pump by analysing the family of curves from different manufacturers and suit to the requirements. This will avoid trimming of impeller in excess of 7% of the original diameter.

Reduction of frictional losses by selection of low friction pipes: Sometimes it is often observed that even for a small duty application high friction loss pipes are installed. Pumps being continuously operated result in operation at normally high heads due to high frictional losses in the pipes. Selection of low friction pipes like PVC will result in operation of pumps at lower head and thereby resulting in power savings.

Reduction of frictional losses by adequate sizing of pipes: This is by far the most neglected area. For optimum pressure drop in the pipes the velocity in the pipes should be within 1.5 to 2 m/s. It has been noticed in several instances that the velocity in the pipes are in excess of 2 m/s. This results in increased pressure drop thereby resulting in operation of pumps at higher head. The design of the pipes has to be carried out in such a way that the velocity is well within acceptable limits.

Also care has to be taken for not over designing the pipes. The pressure drop will be minimal in this case; though the initial expenditure will be high. Hence an optimum solution has to be found out for minimising pressure drop in the pipe lines.

Downsizing of pumps with low capacity pumps: There are cases in industries where a higher capacity pump is installed for a low application duty. The flow is controlled either through variable speed drive, throttling control or trimming of impellers. All these three methods will result in energy savings, however the accurate and optimum solution is downsizing of the pumps as per flow requirements. Energy savings will be more in down-sizing compared to the other options of flow control.

Providing sufficient net positive suction head: When a pump is to be installed, the net positive suction head (NPSH) has to be estimated accurately. The NPSH available should always be more than NPSH required. If NPSH available is less than NPSH required, problems with respect to cavitation will occur which results in decreased output of the pumps.

Installation of overhead tanks: In industries, potable water requirements are met by installing a dedicated pump. Generally, these pumps are continuously operated irrespective of the requirements. This method of operation not only results in loss of energy but also water loss. Both energy and water losses can be minimised by installing an overhead tank. The pumps can then be operated by providing a level controller. Based on the level of the overhead tank required, programming can be done. This results in minimum hours of operation of the pumps thereby resulting in energy savings.

Matching of correct sized motor for the pumps: The motor coupled with the pump has to be sized as close to the brake horse power required to drive the pump for a given application. Installing high capacity motors for low capacity applications will result in operation of motors at low efficiency. Motor efficiency will be reduced if loading is less than 50%. The motors selected should be such that the loading is in the range of 80% to 100%. If motors are loaded more than 100%, the factor of safety as specified by the manufacturer has to be analysed clearly. If motors exceed the factor of safety, it leads to motor failures.

17.3 Parametric approach to energy conservation in pumping

Basically, energy consumption is measured in kilowatt-hours, kWh. Obviously energy consumption can be reduced either by reducing kilowatts or by reducing hours or by reducing both.

17.3.1 Reducing hours

In the case of non-stop, continuous duty processes, e.g., power generation, there is no question of reducing hours. In case of pumping for water supply in a building, where total quantity to be pumped in a day is known, hours can be reduced by pumping at higher rate of flow. This can also be an option for reducing energy consumption, since efficiency of pump with higher rate of flow can often be better.

17.3.2 Reducing kW and overall efficiency

Reducing kilowatts means reducing power drawn. In pumping, power from the supply is drawn, first by the driver. Driver supplies it to the pump after its own efficiency, h_{motor}. Power supplied by the driver to the pump is the power input to the pump. The pump imparts the power to the liquid after its own efficiency h_{pump}. Actually, if the driver is not directly coupled to the pump, i.e., if the pump receives power from the driver through a transmission gear, there will be the efficiency of the transmission gear, h_{trans} which also needs to be looked into. Even when the pump is directly coupled, the transmission efficiency can be poor, if the alignment is not good or if components of the coupling have suffered wear and tear. Thus the right objective for energy conservation in pumping ought to be to improve overall efficiency $h_{overall}$.

17.3.3 Components of overall efficiency

To go into a finer detail, one would notice that supply of power to the driver itself would be happening across an unwarranted length of the cable. Or there can be an unwarranted length of cable between the driver and its controller. And the quality of power supply would also influence the energy-consumption. If one would think that this is stretching the logic too far, think of the bore well pumps. These pumps have a submersible motor and the supply cable has to reach to the motor terminals all the way down. So, there is the efficiency of power supply and control.

17.3.4 Bench-mark information and standards on pumps

When it comes to assessing, what the level of energy consumption should be in a given pumping system, one needs to have information on 'Bench-mark' values. Unfortunately, bench-mark values of energy consumption are not available directly in any standard or in any reference book on pumps. What is available is information on efficiencies of pumps.

Standards on motors also specify efficiencies of motors. In recent years there have been initiatives around the world to enhance efficiencies of motors. International specification of induction motor in U.S. are IEEE 112-B and Germany IEC 34-2 and Indian specification are IS: 13529. In India also there is Indian Standard IS-12516 for energy efficient motors.

17.4   Tips to save energy on pumping systems

1. Select the most efficient pump type for the application: A Finnish research study shows that the average pump efficiency is below 40% and that 10% of pumps are 10% efficient or less. Oversizing often comes in the design phase, since the practice for adding multiple safety factors is quite common. This means that both pressure and flow parameters for the pump design may be 25% more than the actual system operation. The specifying engineer may need to work closely with the pump manufacturer or distributor to optimally select the pump, in addition to its size, speed, power requirements and type of drive, as well as the mechanical seal and ancillary equipment.

2. Proper pump sizing: Proper pump sizing represents a significant economic opportunity to reduce energy consumption. This is important because centrifugal pumps can consume up to 60% of motor energy in a facility and also have the highest process equipment maintenance cost. When engineers add too much of a safety factor during the design phase, the pump can be oversized, resulting in higher energy and maintenance costs.

3. Impeller trimming: The impeller should not be trimmed any smaller than the minimum diameter shown on the manufacturer's pump curve. This is typically about 75% of a pump's maximum impeller diameter. Pump curves and affinity rules (which are valid for a maximum of approximately 5% change in diameter) can both provide information on impeller trim changes and the affected performance. In practice, impeller trimming is typically used to avoid throttling losses associated with control valves.

4. Minimise system pressure drop: A key way to reduce pressure drop is through pipe-sizing optimisation. Hydraulic friction loss creates a reduction in pressure from one end of a straight pipe to another. Factors such as the flow rate, pipe size (diameter), overall pipe length, pipe characteristics (surface roughness, material, etc.), and properties of the fluid being pumped all influence the system pressure drop.

5. Implement proper control valves: Control valves are typically used to control flow and/or pressure. They can help to reduce energy losses over non-controlled systems such as irrigation systems with a fixed-speed pump and multiple locations with different distances and elevations. The main functions of control valves are throttling flow or for bypassing flow. Throttling reduces the flow but increases the pressure. You can minimise excess pressure by bypassing excess flow back to the reservoir or another location.

6. Adjustable speed drives (ASDs): ASDs better match speed to load requirements for pumps. As for motors, energy use of pumps is approximately proportional to the cube of the flow rate and relatively small reductions in flow may yield significant energy savings. New installations may result in short payback periods. In addition, the installation of ASDs improves overall productivity, control and product quality and reduces wear on equipment, thereby reducing future maintenance costs. Similar to being able to adjust load in motor systems, including modulation features with pumps is estimated to save between 20% and 50% of pump energy consumption, at relatively short payback periods, depending on application, pump size, load and load variation.

7. Maintain pumping systems effectively: Effective pump maintenance allows facilities to keep their pumps operating efficiently. Regular maintenance may reveal deteriorations in efficiency and capacity, which can occur long before a pump fails. Wear ring and rotor erosion, for example, can be costly problems that reduce efficiency by 10% or more. Most maintenance activities can be classified as either preventive or predictive. Preventive maintenance addresses routine system needs such as lubrication, periodic adjustments and removal of contaminants. Predictive maintenance focuses on tests and inspections that detect deteriorating conditions. Sometimes called 'condition assessment' or 'condition monitoring,' it has become easier to conduct with modern testing methods and equipment. This can help minimise unplanned equipment outages, which can be very costly.

8. Use higher efficiency/proper pump seals: Sealing systems impact efficiency and mechanical friction losses are only the beginning. Leaks from static and dynamic seals waste fluid can contaminate the environment. Leaks between the pump suction to the pump discharge reduce pump volumetric efficiency. Dynamic seals consume energy from the mechanical friction between the static and moving parts. Potential sealing system savings can exceed the energy savings obtained from switching to variable frequency drives, trimming impellers, or resizing pumps in many applications.

9. Multiple pumps for varying loads: When multiple pumps operate as part of a parallel pumping system, there are opportunities for significant energy savings. A multiple pump parallel system works best when each pump is run individually, not concurrently, most or all of the time. Running multiple pumps simultaneously is appropriate as dictated by the flow requirements specific to the application and duty cycle.

10. Eliminate unnecessary uses: One of the most simple, but often over looked, measures to save energy is to eliminate unnecessary use. Pumping system efficiency measures include shutting down unnecessary pumps and using pressure switches to control the number of pumps in service when flow-rate requirements vary. Each pump system is different and there are many opportunities to save energy.

11. Avoiding throttling valves: Variable speed drives or on-off regulated systems always save energy compared to throttling valves. The use of these valves should therefore be avoided. Extensive use of throttling valves or bypass loops may be an indication of an oversized pump.

12. Replacement of belt drives: Most pumps are directly driven. However, some pumps use standard V-belts which tend to stretch, slip, bend and compress, which lead to a loss of efficiency. Replacing standard V-belts with cog belts can save energy and money, even as a retrofit. It is even better to replace the pump by a direct driven system, resulting in increased savings of up to 8% of pumping systems energy use with payback periods as short as 6 months.

13. Precision castings, surface coatings or polishing: The use of castings, coatings or polishing reduces surface roughness that in turn, increases energy-efficiency. It may also help maintain efficiency over time. This measure is more effective on smaller pumps.

14. Improvement of sealing: Seal failure accounts for up to 70% of pump failures in many applications. The sealing arrangements on pumps will contribute to the power absorbed. Often the use of gas barrier seals, balanced seals and no-contacting labyrinth seals can help to optimise pump efficiency.

17.5 Energy conservation in V-belts and pipe belt conveyors

17.5.1 Energy conservation in V-belts

About one-third of the electric motors in the industrial and commercial sectors use belt drives. Belt drives provide flexibility in the positioning of the motor relative to the load. Pulleys (sheaves) of varying diameters allow the speed of the driven equipment to be increased or decreased. A properly designed belt transmission system provides high efficiency, low noise, does not require lubrication and presents low maintenance requirements. However, certain types of belts are more efficient than others, offering potential energy cost savings.

The majority of belt drives use V-belts. V-belts use a trapezoidal cross section to create a wedging action on the pulleys to increase friction and the belt's power transfer capability. Joined or multiple belts are specified for heavy loads. V-belt drives can have a peak efficiency of 95% to 98% at the time of installation. Efficiency is also dependent on pulley size, driven torque, under or over-belting and V-belt design and construction. Efficiency deteriorates by as much as 5% (to a nominal efficiency of 93%) over time if slippage occurs because the belt is not periodically retensioned.

Cogged belts have slots that run perpendicular to the belt's length. The slots reduce the belt's bending resistance. Cogged belts can be used with the same pulleys as equivalently rated V-belts. They run cooler, last longer and have an efficiency that is about 2% higher than that of standard V-belts.

Synchronous belts (also called timing, positive-drive, or high torque drive belts) are toothed and require the installation of mating toothed-drive sprockets. Synchronous belts offer an efficiency of about 98% and maintain that efficiency over a wide load range. In contrast, V-belts have a sharp reduction in efficiency at high torque due to increasing slippage. Synchronous belts require less maintenance and retensioning, operate in wet and oily environments and run slip-free. But, synchronous belts are noisy, unsuitable for shock loads and transfer vibrations.

Energy savings from synchronous belt drives

Approximately one-third of the electric motors in the industrial and commercial sectors use belt drives. If the efficiency of these systems were improved by a mere 5%, the plants would see tremendous energy savings. Such savings are not out of reach. Synchronous belt drives operate so efficiently that they enable savings across a variety of industrial applications.

Comparing V-belts and synchronous belts: Most of today's belt drives use standard V-belts, which have a trapezoidal cross section creating a wedging action on the pulleys.

Certain physical characteristics of these V-belts cause energy loss. Energy losses in belt drives are separated into two categories, torque loss and speed loss. One factor impacting torque loss is heat generated due to the friction between the belt sidewall and the groove surface of the metal. V-belts depend on friction as they are part of a wedging mechanical system and therefore have greater energy loss due to heat generation than a synchronous drive, which has positive engagement between the belt tooth and sprocket groove and is generally cooler running. Another form of torque loss comes from the energy required to bend a belt around a sprocket or sheave. The thinner cross section of a synchronous belt requires less energy to bend than the thicker cross section of a V-belt. Figure 17.1 shows comparison of a synchronous belt drive and V-belt drive.

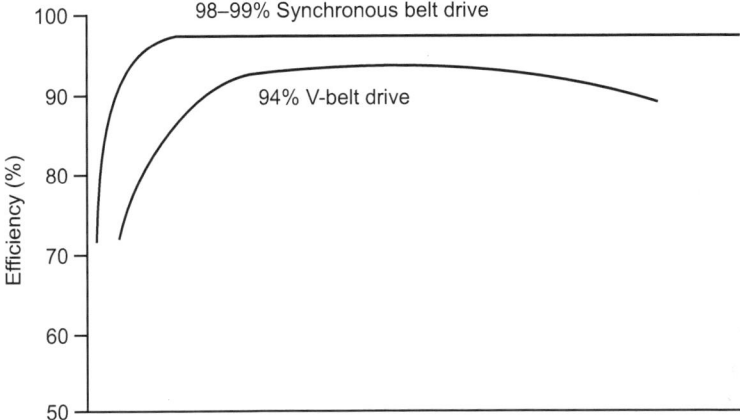

Figure 17.1: On average, a synchronous belt drive is 5% more efficient than a V-belt drive, eliminating excess energy consumption.

Speed loss is also a characteristic of V-belt drives. A positive tooth/groove engagement prevents a synchronous belt drive from slipping, while V-belt drives, no matter how well maintained, will exhibit some amount of slip. Slip occurs when the tension is insufficient to transmit the load. V-belts elongate and require retensioning on a regular basis whereas synchronous belts have minimal elongation and require no retensioning if properly installed.

Although properly maintained V-belt drives can run as high as 95–98% efficient at the time of installation, this deteriorates as much as 5% during operation. Poorly maintained V-belt drives may be up to 10% less efficient.

Synchronous belts, remain at an energy efficiency of approximately 98–99% over the life of the belt.

Increased efficiency equates to savings

To determine the kilowatt-hours saved when using synchronous belt drives rather than V-belt drives, the following formula is used:

$$kWh = \frac{(\text{Motor HP})(\text{Hrs/Yr})(0.746)(0.05)}{\text{Motor efficiency}}$$

where, constant 0.746 is the conversion factor from hp to kW and 0.05 is the 5% energy savings gained by converting.

Maintenance managers can leverage improved energy efficiency by converting V-belt drives to synchronous belt drives in one of two ways:

1. Maintaining current capacity while using less power.
2. Increasing capacity using the same power.

For example, if the current airflow is satisfactory in an HVAC application, a synchronous belt drive would use less energy to do the job. If the current airflow is insufficient, a synchronous belt drive could increase airflow without increasing use of energy.

Saving maintenance expense and downtime: V-belt drives and synchronous belt drives demand approximately the same amount of time for installation. A key difference between them, in terms of maintenance, is that synchronous belts do not require a run-in procedure or retensioning. It is recommended that a newly installed V-belt is retensioned 24 hr after installation. Time is spent locking out the power, removing the belt guard, retensioning, securing the belt guard and resuming power. Companies that are too busy to do the proper 30 minute run-in process are later burdened by premature belt failure. That means costly belt replacement is needed. In addition, V-belts should be retensioned based on a scheduled preventive maintenance programme for optimum performance. Like run-in, each procedure takes approximately 30 minutes, during which the drive must be shut down and productivity is lost. On critical drives, a synchronous belt, which requires no retensioning, not only improves energy efficiency but also eliminates downtime. More uptime equates to more production, which leads to higher profit.

Quick facts

1. Industrial motor use consumes 25% of total electricity usage in the U.S.
2. The majority of belt failures can be traced to environmental factors (debris, temperature, contaminants) and improper belt drive maintenance.

V-belts and maintenance

Most of today's belt drives use standard V-belts, which have a trapezoidal cross section creating a wedging action on the pulleys. V-belts depend on friction as they are part of a wedging mechanical system.

V-belt drives can run as high as 95–98% efficient at the time of installation. They are manufactured in a wide variety of materials, cross sections, banded multiples, reinforcement styles and constant and variable speed configurations. Low acquisition costs, wide availability and quiet performance make them a popular power transmission solution.

A key difference between V-belts and synchronous, in terms of maintenance is that synchronous belts do not require a run-in procedure or retensioning. It is recommended that a newly installed V-belt is retensioned 24 hr after installation. V-belt drives and synchronous belt drives demand approximately the same amount of time for installation.

Additional tips include:

1. Don't store belts near chemicals, oils, solvents, lubricants, or acids. Exposure to these potentially harmful chemicals can reduce belt performance and overall life. Gates recommends replacing sprockets after every third replacement belt has reached its maximum service life, or sooner if the sprockets show significant wear. If the sprockets show significant wear, the life of the next replacement belt will be considerably reduced.

2. If belt drives operate under conditions of high speeds, heavy loads, frequent starts and stops and temperature extremes or on critical machinery, check them for unusual noise, vibration, or visual deterioration more frequently. Depending on the critical nature of the drive, this monitoring could be necessary as frequently as every 1 to 2 weeks.

3. Improperly tensioned drives will experience shorter life. Belt tension should not be too high or too low. Tension that is too low can also lead to shortened belt life or synchronous belt ratcheting, while tension that is too high adds undue stress to bearings, shafts and other related components.

4. Increasing ventilation around belt drives can help reduce belt operating temperatures in applications with high ambient temperatures. This can be accomplished by adding vents to belt guards, by providing a cooler external air source, or even by adding fins to sheaves.

5. Never pry or roll belts onto sheaves. This will cause invisible damage to the tensile cords and reduce belt life. Belts should be stored on a flat surface where possible. Hanging belts on hooks can cause belt crimping and lead to reduced belt life, especially with larger and heavier belts.

17.5.2   Pipe conveyor belt

Pipe conveyor is a modern and environmentally friendly transport system solving numerous problems associated with conventional conveyor system, i.e., spillage of materials, limitations with regard to steep incline and curve layout, etc.

Pipe conveyor also manages to transport difficult materials, i.e., Powder and/or materials that are similar to liquid. Bridgestone is the pioneer of the pipe conveyor system since acquiring patents and rights from Japan pipe conveyor co. ltd., in 1987 and has around 1000 supply records worldwide since the 1980's.

Energy efficient operation of long overland pipe conveyors

Although the same is true for trough conveyors, avoiding empty running for pipe conveyors has a stronger effect on the operating cost. With variable frequency drives, the power consumption can be reduced by adjusting the belt speed, depending on the material load, while keeping the same material cross section. Keeping the material cross section constant can also increase the pipe belt stability and reduce belt rotation. For long overland pipe conveyors, this can be achieved with a silo or stockpile at the tail, weight scales near the loading point and Variable frequency drives programmable logic controllers (VFD's PLC) control logic to adjust belt speed based on material loading.

Basic advantages of pipe conveyors

Environmental protection and totally enclosed conveying: One of the big advantages of a pipe conveyor is that the material transported is contained within the rolled pipe/tubular shape of the belt for the majority of its conveying distance.

This has the following benefits:

1. Environmental pollution is minimised as enroute spillage is eliminated.
2. The material conveyed is protected from losses due to wind, spillage, contamination, rain and theft.

High angle conveying: The increased friction between material and the pipe shape of conveyors due to packing makes generally a 50% higher angle of conveyance possible by pipe conveyors as compared with conventional toughed belt conveyors to angles of vertical inclination as high as 30°.

This results in:

1. Smaller space required for installation, making pipe conveyors viable solutions if there are space restrictions within the plant.
2. With the steeper angle of inclination, the overall length of the conveyor system can be reduced.

Complex 3D profiles: The flexibility of the pipe shape permits the belt to be curved both horizontally as well as vertically. In many instances this is a big advantage as it eliminates transfer points where there is a relatively sharp change in conveyor direction. A single conveyor can thus replace several conventional belt conveyors reducing:

1. The need of multiple transfer point, drives, dust collection systems, structures, etc., all prone to operation and maintenance issues.
2. Reducing particle degradation resulting from transfer point.
3. This makes the system compact and saves capital investment.

Return belt conveying: The basic design feature of the pipe conveyor belt enables using the return side of the belt for conveying materials in the opposite direction. The conveying in of materials and conveying out of materials in a single belt is possible in longer belt conveyor installations especially for plants in the vicinity of ports that offer distinct cost advantages.

Same volume of material transported: A pipe conveyor transports the same volume of material that a conventional troughed belt conveyor, 2.5 to 3 times its pipe diameter transports. This means that pipe conveyor requires support structures of narrower widths and often lesser weights.

Power saving: When a single pipe conveyor replaces several conveyors in a system, the total power consumed is a considerably lower. A series of several conventional belt conveyors also require additional power to lift material at each transfer point.

Maintenance friendly: As a single pipe conveyor replaces several trough belt conveyors, it allows the use of travelling maintenance trolleys, providing easier maintenance. It saves a lot a steel and other materials required in conventional conveyors and therefore the cost.

Energy efficient fuel oils and lubricants in textile industry

18.1 Introduction

Fuel oil is scarce, expensive and has better uses than firing in an utility boiler. But for the recent developments, fuel oil has been traditionally fired even in coal fired boilers, for ignition and stabilisation of the coal flame. By understanding and careful planning, control and maintenance of the system and equipments, the fuel oil can be greatly economised.

18.2 Conservation of fuel oils in textile industry

18.2.1 Storage, handling and preparation of fuel oils

Furnace oil as delivered may contain dirt, gummy substance and water. During storage for prolonged period, above impurities tend to form sludge which accumulates in the various dead spaces of the tank. This results in clogging of filters and unsteady burner flame ultimately. Cleaning of main storage tank, about once in a year, is recommended to remove the accumulated sludge. Also, the furnace oil supplied may contain some water in the form of dissolved moisture or dispersed free water.

After receiving the product in main storage tank, settling time of approximately 24 hr should be given for the dispersed free water to settle down at the tank bottom which should be drained from the tank. Once in a day water drainage schedule should be maintained to keep the water content in the storage tank at minimum level.

18.2.2 Maintaining proper oil pre-heat temperature

Maintaining correct oil viscosity at burner tip ensures proper atomisation of oil and this results in efficient combustion operation. However, due to varying characteristics of the fuel oil from one batch to another, maintaining a constant oil pre-heat temperature does not serve any purpose. Hence either an on-line viscometer should be installed near the burner which automatically adjusts pre-heat temperature to maintain the recommended viscosity at burner tip or sample should be drawn every time from the service tank to check the viscosity and accordingly pre-heat temperature should be adjusted.

18.2.3 Requirements for efficient firing of fuel oil

1. Preparation of the fuel for combustion which consists of pre-heating the fuel to the required temperature to reduce viscosity and atomising this fuel into fine particles for exposure to maximum surface area for rapid conversion of liquid fuel into gas.

2. Conversion of the complex fuel into elementary fuels (gases): The speed with which the atomised fuel vapourises depends on the size of the droplets, the temperature in combustion space and speed of vapour removal by combustion air which again depends on the relative velocity between liquid oil particles and air.

3. Bringing these fuels and air together in the right proportions and at the proper time and temperature for ignition and combustion. To ensure complete combustion, more air is necessary than is theoretically required. The smaller the percentage of excess air used to obtain complete combustion in the boiler, the more efficient is the installation. The efficiency of combustion may be considered very good if the excess air is kept below 22% and is excellent if it is below 10% at the time of firing fuel oil. The excess air is an important factor for achieving efficient combustion and consequent fuel economy in boiler operations.

18.2.4 Stoichiometric combustion of fuel oil in boilers/thermic fluid heaters

The stoichiometric combustion for 1 kg of fuel oil requires 14.1 kg of air and carbon is fully converted to carbon dioxide releasing complete heat of combustion. The carbon dioxide constitutes 15.9% by volume of flue gases and there is no oxygen in flue gases.

In actual practice, since mixing of fuel and air is never perfect, a certain amount of excess air is needed for complete combustion and to ensure the release of the entire heat contained in the fuel oil. If air is too high than what is required for complete combustion, then this excess air gets heated up and leaves the boiler constituting high stack losses, seen by colourless smoke at the chimney. On the other hand, if quantum of air is not sufficient then it would lead to incomplete combustion of fuel, seen by black smoke.

For proper combustion, optimum excess air level, normally 10–20%, should be maintained which would be consuming the least fuel and giving the maximum boiler efficiency.

The excess air can be controlled by monitoring oxygen or carbon dioxide content in the flue gases. It is recommended to use an electronic flue gas analyser to monitor O_2 and CO_2 level in flue gas once in every shift.

18.2.5   Efficient utilisation of steam

Nearly 66% of fuel oil used in industry is for generating steam in industrial boilers. The very purpose of generating steam at the highest thermal efficiency is lost if the steam is subsequently wasted through improper distribution or utilisation. Efficient steam utilisation not only results in saving of fuel oils, but also reduction in the peak demand for steam bringing it within operating capacity of the boiler house, thereby avoiding fuel wastage through overloading or the need for additional boilers. Overall production time too could be reduced-thus improving the productivity of the steam consuming manufacturing processes. Following measures may be adopted for efficient utilisation and distribution of steam.

Avoiding steam leakages

Steam leakage is a visible indicator of steam waste and must be avoided. It has been estimated that a 3 mm diameter hole on a pipeline carrying 7 kg/sq. cm pressure steam would waste 32.65 kL of fuel oil per year. Steam leaks on high pressure mains are very high compared to that on low pressure mains. By plugging all leakages at pipelines, valves, flanges and joints, fuel savings corresponding upto 5% steam consumption in a small or medium scale plant can be achieved.

Provision of dry steam for process

Wet steam carry no latent heat, reduces the total heat in the system and also increases the water film on heat transfer surfaces reducing rate of heat transfer. Moisture in wet steam also overloads the traps and other condensate handling equipments. On the other hand, superheated steam is also not desirable as its temperature can not be effectively controlled and it releases its heat at a rate slower than the condensate heat transfer or saturated steam. Hence, best steam for industrial process heating is dry saturated steam. The dryness fraction of steam depends on factors such as water level in boiler drum, effects of peak loads, the surging within the boiler, the pressure on the water surface in the boiler and the solid content in the boiler water. The extent of progressive condensation of steam, as it flows through the pipelines, depends on the effectiveness of the pipeline lagging. A steam separator may be installed on the steam mains and branch lines to reduce wetness in steam for improving the steam quality going to the user units.

Steam utilisation at lowest practicable pressure

Latent heat in steam, which takes part in process heating in an indirect heating system, reduces with increase in steam pressure. It is advisable to generate and distribute steam at highest possible pressure but to utilise at as low a pressure as possible since it then carries higher latent heat. But decrease in steam pressure happen with drop in temperature which, in turn, slows down heat transfer rate thereby increasing processing time thus steam consumption in equipments with high fixed losses. Therefore depending on the equipment design the lowest possible steam pressure, with which the equipment can work, should be selected without sacrificing either on production time or on steam consumption.

Insulation of steam pipelines and hot process equipment

It has been estimated that a bare steam pipe, 150 mm in diameter and 100 metres in length, carrying saturated steam at 8 kg/sq.cm, could waste 25 kL of furnace oil in one year due to heat loss to atmosphere by radiation. In case, the pipelines are already lagged, the effectiveness of the insulation must be reviewed periodically.

18.3   Conservation of lubricants in textile industry

18.3.1   Energy saving through spindle oil in ring frame

The Indian textile industry is one of the oldest and largest among all the industries and also the second largest in the world. As per an estimate, around 29 million ring spindles are in operation in India. It is estimated that the ring frame section consumes more than 70% of the total electrical power in a spinning mill. Out of this about 40% is consumed to drive the ring spindles. It is assumed that a saving of Rs. 20 crore could be achieved per annum by the use of a proper spindle oil giving a moderate energy saving efficiency of 4%. Amongst the various textile oils, the spindle oil, i.e., the lubricant used in ring frame has attained a special significance in the light of the above reason.

18.3.2   Power consumed in ring spinning

By looking into the break-up of power consumption in various departments of a spinning mill, it is reported that around 80% of power is consumed by ring frame alone. Other major power consuming departments are cards followed by speed frame and comber.

18.3.3   Optimum oil level in the spindle bolster

The power consumption increases with increase in oil level in the bolster because of resistance offered by the oil. Also, excessive oil level in the bolster may disturb the proper running of the spindle. Usually, 3/4th of the bolster capacity is filled with oil. The normal mill method of determining the depth of oil level in the bolster is by lifting the spindle and observing the oiliness on the spindle blade.

This way of assessment may not be so accurate to find out the normal and minimum level of oil required. The correct or exact amount of oil for each type of spindle insert could be ascertained by using dipstick. The dipstick has two distinct markings, i.e., the bottom marking for minimum and the top marking for maximum oil levels. The normal tendency in many mills is to fill up the bolster to a near full level at the time of filling and topping up which leads not only to wastage of lubricant but also to higher power consumption. The oil level should therefore be checked with a dipstick after every topping and the level should be maintained within the limits prescribed by the machinery manufactures. In this regard various oil dosing equipments are available today for filling and topping spindle bolsters with the pre-determined quantity of oil. The novel features of these equipment are their volume control mechanism, which facilitates to dose out required quantity of spindle oil (5 cc to 20 cc) to suit the capacity of different bolsters used in ring frames. As there is no excess filling of spindle oil to the bolster, it prevents wastage of oil as well saves in energy.

18.3.4   Spindle oil and its essential characteristics

Amongst the various textile lubricants, ring frame consumes large quantity of spindle oil. The ring spindles are run in the speed range of 10,000–20,000 rpm. The spindles operate both in hydrodynamic and boundary lubrication conditions. Most of the spindle oils used in the industry are in the viscosity range of 9.0 to 11.0 centistokes (cST) at 40°C, i.e., ISO VG 12 grade as per IS 493 (Part I): 1993 for spindle oils. Usually ISO VG 12 grade spindle oil is recommended for ring spinning and ISO VG 22 grade for heavy doubling machines. Spindle oils used in ring frame have attained a special significance in the light of energy saving. In high speed spindles, the spindles operate both in hydrodynamic and boundary lubrication conditions.

Energy savings: Energy savings can be achieved through optimum viscometrics of the spindle oil and by incorporating certain dispersant additives along with the base spindle oil. Oils with relatively lower viscosity is found to be energy conserving. Therefore, it is seen that the viscosity has a great role in energy saving. The dispersant additive gives an effective dispersion of the sludge generated in the spindle bolster of a very small size and thereby providing superior lubrication. Spindle oil with higher viscosity index undergoes lower variations in viscosity with temperature and its lower evaporation loss is beneficial in reducing oil make-up costs. These two factors are therefore also important for energy saving.

18.3.5   Use of energy efficient spindle oils

The incorporation of dispersant additive system to the mineral-based spindle oil may give power savings up to 3% when compared to conventional oil. The amount of actual saving will depend upon the condition of the machinery and their operation. Energy saving also can be achieved by using light weight spindles.

However, synthetic-based spindle oils (energy efficient grades or EE grades) along with certain metal compatibility additives, etc., may give a higher power saving in the range of 5–7% depending upon viscosity. For monitoring the service life of these additive added oils, the total acid number (TAN) build-up, metal compatibility (copper and steel) and sludge forming tendency are to be examined. An alarming increase in TAN, very high amount of sludge and poor copper strip corrosion are the indication of poor durability of the oil. The energy efficiency of a particular oil can be assessed in two ways, i.e., (i) power consumption in kWh and (ii) reduction in bolster temperature rise over ambient: energy savings oils show less temperature on use.

While selecting any energy saving spindle oil, the important characteristics related to the service life of oil, i.e., temperature rise, thermal stability, metal compatibility, sludge forming tendency and anti-wear/antifriction properties have to be evaluated. The energy saving in a spinning mill can be achieved through the following measures:

1. Fill only optimum quantity of oil to the bolster, i.e., do not fill-up excess quantity which may increase power consumption.
2. Use of light weight spindles is vital to reduce friction, thereby leading to saving in power. Now-a-days conventional cotton tapes for driving spindles have been replaced by synthetic spindle tapes. This may also save power consumption.
3. Use energy saving spindle oil after proper checking for its important properties.

All these measures are expected to give power saving as well as trouble free service, equipment protection, reduced downtime, etc. By way of saving 3–6% energy the Indian textile spinning units could save 48–96 million units per annum, i.e., savings of Rs 20–22 crores in electricity bill.

18.3.6 Conservation of lubricants through proper storage and handling practices

With proper implementation of correct handling and storage procedures, a saving of at least 3% by way of lubrication cost, both direct and indirect, in terms of lubricant life, leakage, machine downtime, etc., may be achievable.

Moreover, the benefits to be derived through the use of high quality lubricating oils and greases can be easily offset if proper care is not taken during storage and handling of lubrications. Any contamination with dirt, dust or moisture will impair the lubricating quality of the products and may lead to excessive wear, loss of power and early breakdown. Some guidelines on storage and handling of lubricants and greases, which, if implemented, will help to prolong the life of the lubricants and also ensure protection to plant and equipment through the use of proper quality of lubricants are given below.

18.3.7 Storage

Lubricating oils

1. Oil barrels should preferably be stored indoors, in a covered storage area, horizontally with the bungs in the 3 O'clock –9 O'clock position only to prevent ingress of moisture from the atmosphere through breathing action through the bungs.

2. Oils of different grades should be stored in separate lots with a board hung near each lot indicating the name of oil. Separate stacks should have sufficient spacing to facilitate handling and movement of barrels.

3. Where outdoor storage is unavoidable, the barrels should be stored horizontally, with the bungs in 3 O'clock–9 O'clock position only, on wooden battens. Water logging should not be allowed in the barrel storage area.

4. In case any barrel has to be stored on its end, it should be kept tilted so that rain water does not collect on the top of the barrel.

Greases

1. Grease barrels should be stored vertically in covered storage since they are sensitive to contamination with moisture and extraneous materials.

2. Wherever outdoor storage is unavoidable, the barrels should be kept separated from the ground with help of wooden dunnage and covered with tarpaulin to prevent water collection on the lid.

Handling

1. After opening a barrel the bungs/lid should be kept closed to prevent impurities, dust, water or fibre from entering. Such contaminants eventually finding their way into machinery can cause damage by abrasion or by blocking oil passages leading to a complete breakdown due to lack of lubrication.

2. Dispensing contractors should not be dipped into lube oil/grease drums. Wherever possible drums of lube oil in use, should be placed on wooden cradles of convenient height and oil should be dispensed by means of a tap with a drip tray placed underneath. Alternately, the drum may be stored on its end and product withdrawn with a clean hand pump.

3. Containers used for transporting lubricants within the factory must be kept clean and provided with lids. Use of cotton waste or woolen rags should be avoided since they tend to leave behind fibres which may find their way into the machinery.

4. It is advisable to have separate clearly marked dispensing containers for each group of oils and for individual greases.

5. In general, greases are very susceptible to chances of contamination both due to the fact that the entire top lid can be opened and, unlike oils, external contaminants do not sink to the bottom. Special care therefore needs to be taken in handling greases. Clean wooden spatulas should preferably be used for grease dispensation.

Essential characteristics of a good textile spindle oil

The essential characteristics of a good textile spindle oil are: (i) low viscosity, (ii) low reacting acidity, (iii) good rust protection, (iv) good thermal stability, (v) superior antifriction property and (vi) good metal compatibility.

18.3.8 Measures adopted for energy savings

Measures adopted for energy savings are: (i) bolsters must be topped upto the level specified by the manufacturer. Excessive oil leads to increase in power consumption, (ii) use of dipstick is recommended for checking the depth of oil in the bolster, (iii) use ISO-VG-10 grade oil for ring spinning and ISO-VG-22 grade for heavy doubling machines, (iv) use lightweight spindles and (v) incorporate dispersant additive system in the mineral based spindle oil.

18.4   Rationalisation and conservation of lubricants for textile industry

In textile industry, rationalisation of lubricant is very important. Rationalisation means changing over to equivalent or better quality oil of wider application. Every effort must be made to cut down wastage and ensure economic consumption of lubricants. Rationalisation is an important step in achieving conservation of lubricants. Many lubricants of similar properties or viscosities are used in various sections of the mills is given in Table 18.1. Generally, a high inventory of lubricants may result in unorganised and oversized lubricant list which may lead to frequent purchase of oil through retailers in small quantities. This, in turn, could lead to quality problems through the use of substandard lubricants causing machine wear and tear. Similarly, the procedure for procurement of lubricants in the entire mill could save several man hours. In India, the lubricant demand is increasing at the rate of 12% per annum. This aspect has to be viewed seriously in the background of exhaustible natural hydrocarbon resources available to mankind. The Indian textile industry is the second largest in the world. For instance, of the world spindleage of 90 million, India accounts for 30 million spindles.

Table 18.1: Section-wise and machine-wise viscosity range of oil used in textile industry.

Section/machine	Viscosity range in (cst)
Blowroom	12–460
Card	32–460
Comber	12–460
Draw frame	12–460
Speed frame	12–460
Ring frame	12–460
Winding	32–460
Warping and sizing	100–460
Plain loom	100–150
Ordinary auto loom	100–326
Air-jet loom	30–680
Processing	68–460
Other speciality lubricants	
High temperature general purpose machine oil	EP 150–680
Chain lubricant	Synthetic oil - 220
Liquid grease	Draft lube. 6/14
Grease	
General purpose	Multi purpose 2/3 grade (230/295)
Special purpose	High temperature grease > 260°C
Spray	Moly spray (Based on MOS_2)

A recent analysis of the oil extraction and topping frequencies followed in the industry has revealed staggeringly wide variations between the mills. It was also found that a large quantity of extracted oil is either disposed off or burnt in the boiler resulting in poor realisation of the lubricant. In many cases good oil is prematurely removed. Similarly, the schedule of oiling of machine parts also varies from mill to mill. In the case of oil topping, the details collected from various mills also indicate the same trend. During the survey it has been observed that some high productivity mills are following optimised schedules of lubrication. This section highlights the present topping and extraction schedules of various textile machines. Studies have also indicated that highly refined spindle oil and gear oil can be used in the enclosed system for a longer duration without affecting the lubricating properties.

Rationalisation of lubricants does not just mean reducing the number of lubricants haphazardly.

The key to successful lubrication is the selection of proper lubricant which depends on the technical and operational requirements, the individual friction point has to meet. Thus, lubrication and lubricants are considered to be important issues in terms of technology and materials management.

Substantial quantity of lubricant can be saved by:

1. Extending the topping schedule.
2. Avoiding premature removal of good quality oil.
3. Providing good quality oil seal for gear boxes.

18.4.1 Energy efficient lubricants for textile industry

Some of the important lubricants used in the textile industry are:

1. Textile spindle oils.
2. Scourable loom oils
3. Industrial gear oils.
4. DG sets oils.
5. Greases, etc.

However, for the textile industries to remain competitive in the global market, there is a need to improve the quality of the lubricants so as to bring down the cost of operation and eliminate rejections arising from the presence of oil stains on the fabric.

Energy efficient spindle oils

Ring frames consume around 40 to 50% of electrical energy in a composite mill and around 60% in a spinning mill. Out of this 50% of the energy is consumed by the moving spindles.

Energy saving by modifying the spindle system will result in substantial savings for the textile industries and the country.

Energy savings in ring spinning and twisting have been reported through use of spindle oils incorporating synthetics and suitable additive systems. Most of the energy in spindle is dissipated in the bolster and it is independent of count and sort spun. The lower portion of the spindle blade is immersed in spindle oil which provides lubrication to the two bearings, i.e., pivot bearing at the bottom and roller bearing at the neck. The properties of oil play a vital role in the performance of the spindle in terms of life and power consumption.

Properties of spindle oil: The spindles operate both in hydrodynamic and boundary lubrication conditions. Energy savings can be achieved through optimum viscometrics of the spindle oil and by selection of appropriate additive system. The use of dispersant additive in spindle oil formulation was tried on the basis that this additive will give an effective dispersion of the sludge generated in the spindle bolster of very small size and thereby providing lubrication. Based on the above, the formulation of energy efficient oil was finalised, by adopting following approaches:

1. Proper viscometrics.
2. Use of synthetics.
3. Use of friction additives.
4. Combination of above.

Since, spindle oil is exposed to severe conditions during its service such as oxidation, high temperature and contamination with moisture, these lead to deterioration of quality of oil over a period of time.

Therefore following characteristics are considered essential for a good textile spindle oil:

1. Adequate viscosity.
2. Low reacting acidity.
3. Good rust protection.
4. Good thermal stability.
5. Superior antifriction property.
6. Good metal compatibility.

18.4.2   Energy efficient industrial gear oil

Industrial gear oils containing sulphur phosphorous (SP) compounds are recommended for heavy duty enclosed gear drives with circulation and splash systems operating under heavy or shock loads. These oils are also used as EP type oils in plain and roller bearings, sliding surfaces, chain drives, sporockets, flexible couplings employing circulation or spray lubrication system. These oils are in use in variety of industries like cement, steel, power, fertiliser, textile industries, etc. Depending on the operating conditions of the gears, the viscosity of these oils vary from VG 68 to VG 680.

Gears play a vital role in all areas of the textile industry. To make production as efficient as possible, make sure you select the right gear oil for your application. For selecting the right oil for your gears, parameters such as performance, speeds, environmental influences and special operating conditions need to be taken into consideration.

Based on this information, it is possible to select the: (i) oil type, (ii) wear protection and (iii) viscosity which enables the gear oil to optimally perform its tasks, which include: (i) absorbing forces, (ii) reducing friction, (iii) minimising wear, (iv) dissipating heat and (v) absorbing wear and contamination.

Requirements to be met by gear oils

Gear oil properties are determined by the base oil and the additives. The essential requirements for gear oils are described by leading gear manufacturers in international standards and specifications. They include:

1. Operating temperature range.
2. Viscosity.
3. Ageing behaviour.
4. Low-temperature behaviour.
5. Corrosion protection on steel/nonferrous metal.
6. Foam behaviour.
7. Elastomer compatibility.
8. Compatibility with interior coatings.
9. Wear protection–fretting, micropitting.

Elastomer compatibility of gear oils

In the textile industry, compatibility of the gear oils with elastomer is crucial. The materials used for radial shaft seals (RSS) or static seals, e.g., O-ring seals, must not become brittle or softer when exposed to gear oil as their sealing capacity will be affected.

The seals would suffer premature wear, leading to leakage. Cleaning and possibly expensive gear repairs will then be necessary.

Compatibility with the seals should be considered especially when higher torques lead to higher operating temperatures, or when a gearbox is changed from mineral to synthetic oil lubrication. The tests used for verifying the static and dynamic compatibility of gear oils with elastomers are based on ISO 1817 and DIN 3761, respectively.

If the materials to be used are selected carefully, run-times of more than 20,000 hr can be attained.

Normal oil change without change over

Oil-lubricated gears require an oil change from time to time since the oil changes its characteristics beyond limits due to the operating and ambient conditions, e.g., ageing, abrasion and contamination. The objective of the oil change is to ensure continued reliable lubrication. This is also the objective when replacing a gear oil that is basically still fit to use, but not under the prevailing operating conditions.

When an oil change of this type is performed, some residual amount of old oil will always remain in the gearbox. In many cases, these residues cannot be tolerated and some way of removing them must be found. The simplest method is flushing the gears. If possible, the old gear oil is drained while still warm, i.e., immediately after the gears are stopped. With the subsequent flushing procedures, further residues are removed. The oil container and inside walls of the gearbox can also be cleaned using a non-fraying cloth (do not use cleaning wool) and a rubber blade.

Profound contamination in the form of deposits caused by strongly aged oil pose a major challenge. In such cases, cleaning oil must be used and all accessible parts of the gearbox cleaned manually.

Oil change checklist – Gear inspection

Clean gear:

1. Drain oil while warm.
2. Inspect teeth.
3. Replace filters.
4. Fill in new oil.
5. Put gear into operation and stop again.
6. Check oil level.
7. Take reference oil sample, if required.

Contaminated gears:

1. Drain oil while warm.
2. Fill flushing oil.
3. Operate gear for approximately 30 to 60 min without load or injection system only.
4. Drain flushing oil.
5. Inspect teeth.
6. Replace filters.

 7. Fill in new oil.
 8. Put gear into operation and stop again.
 9. Check oil level.
10. Take reference oil sample, if required.

Strongly contaminated gears:

 1. Drain approximately 10% of the oil fill while warm.
 2. Top up with solvent.
 3. Operate gears for 24–48 hr.
 4. Drain oil while warm.
 5. Fill flushing oil, if required.
 6. Operate gear for approximately 30 to 60 min without load or injection system only, if required.
 7. Drain flushing oil, if required.
 8. Inspect teeth.
 9. Replace filters.
10. Fill in new oil.
11. Put gear into operation and stop again.
12. Check oil level.
13. Take reference oil sample, if required.

Rolling bearing

They may be designed as ball or roller bearings radial or thrust bearings: What they all have in common is the transmission of load and power via rolling elements located between bearing rings. This is a simple and successful principle, at least as long as the contact surfaces remain separated. However, if the surfaces come into contact with one another, there can be trouble ahead: the resulting damage caused may be anything from light, hardly perceptible surface roughening, pronounced sliding and scratching marks, to extensive material transfer that may promote premature bearing failure – with expensive consequences! A vital requirement for low-wear or even wear free operation of rolling bearings is the sustained separation of the surfaces of rolling elements and raceways, i.e., the friction bodies, by means of a suitable lubricant. Ideally, it should wet all the surfaces in the bearing.

Grease application in rolling bearings

Around 90% of all rolling bearings are lubricated with grease. Grease lubrication presents far fewer sealing problems than oil lubrication and allows much simpler machine designs. With grease-lubricated rolling bearings we differentiate between lifetime lubrication and bearings which require relubrication. In general terms lifetime lubrication does not depend on the bearing but on the requirements of the particular application.

Relubrication – compatibility of greases

The issue is whether the new grease really is compatible with the old one. Compatibility should be checked with great care.

If the two greases are incompatible, liquefaction, overheating or bearing damage can be the consequence.

Right amount of grease in the bearing

The correct quantity of grease will vary based on the bearing type and bearing rotational speed. It is therefore important to determine the precise grease quantity for the bearings prior to change over. Purging of a bearing with fresh grease will involve completely filling the bearing with grease. This method may prove unsuitable, for instance, when considering high-speed bearings which require an extremely low percentage of grease fill.

Some benefits are: (i) less down time, (ii) higher production and (iii) low oil top up rate.

18.4.3 Planned lubrication programme

A planned lubrication programme includes the following aspects:

1. Choice of proper lubricants for various applications.
2. Rationalisation of the number of lubricants.
3. Identification of lubricants through proper colour coding on the containers.
4. Method of lubrication through use of proper contrivances.
5. Scheduling of the lubrication.
6. Storage and handling of lubricants.
7. Conducting lubrication survey and establishing the norms of lubricants consumption.
8. Analysing the consumption reports and excercising control.
9. Investigating special problems and training the personnel.

Main factors for selection of lubricants

While selecting the lubricants, following main factors should be given due importance:

1. Speed.
   (a) For lower speed—grease preferred.
   (b) For higher speed—oil preferred.

2. Load.
3. Temperature.
4. Reliability.

Guidelines for selection of lubricants

On the basis of the above, following guidelines should be followed for selection of any lubricant:

Condition	Possible choice
Higher load	More viscous oil
Lighter load	Lighter oil
Lower speed	Heavy oil
Higher speed	Lighter oil with more circulation
High temperature	Viscous oil or grease with molybdenum disulphide
Low temperature	Dry lubricants or less viscous oil

Energy saving in cooling towers

19.1   Introduction

A cooling tower is a heat rejection device which rejects waste heat to the atmosphere through the cooling of a water stream to a lower temperature. Cooling towers are a very important part of many industrial plants. Cooling towers may either use the evaporation of water to remove process heat and cool the working fluid to near the wet-bulb air temperature or, in the case of closed circuit dry cooling towers, rely solely on air to cool the working fluid to near the dry-bulb air temperature.

Cooling towers are an integral component of many refrigeration systems, providing comfort or process cooling across a broad range of applications. They are the point in the system where heat is dissipated to the atmosphere through the evaporative process and are common in industries such as textile industry, oil refining, chemical processing, power plants, steel mills and many different manufacturing processes where process cooling is required.

They are also commonly used to provide comfort cooling for large commercial buildings including airports, office buildings, conference centers, hospitals and hotels.

Cooling tower structures vary greatly in size and design, but they all function to provide the same thing: liberation of waste heat extracted from a process or building system through evaporation of water. In technical terms, cooling towers are engineered and designed based on a specified cooling load, expressed in refrigeration tons. The cooling load is determined by the amount of heat that needs to be extracted from a given process or peak comfort cooling demand. The cooling tower must be adequately sized to reject this same amount of heat to the atmosphere.

Cooling towers are used to reject heat through the natural process of evaporation. Warm recirculating water is sent to the cooling tower where a portion of the water is evaporated into the air passing through the tower. As the water evaporates, the air absorbs heat, which lowers the temperature of the remaining water. This process provides significant cooling to the remaining water stream that collects in the tower basin where it can be pumped back into the system to extract more process or building heat, thereby allowing much of the water to be used repeatedly to meet the cooling demand. The amount of heat that can be rejected from the water to the air is directly tied to the relative humidity of the air. Air with a lower relative humidity has a greater ability to absorb water through evaporation than air with a higher relative humidity, simply because there is less water in the air. As an example, consider cooling towers in two different locations–one in Atlanta, Georgia and another in Albuquerque, New Mexico. The ambient air temperature at these two locations may be similar, but the relative humidity in Albuquerque on average will be much lower than that of Atlanta's. Therefore, the cooling tower in Albuquerque will be able to extract more process or building heat and will run at a cooler temperature because the dry desert air has a greater capacity to absorb the warm water. Cooling towers can be split into two distinct categories: open circuit (direct contact) and closed circuit (indirect) systems.

In open circuit systems the recirculating water returns to the tower after gathering heat and is distributed across the tower where the water is in direct contact with the atmosphere as it recirculates across the tower structure. Closed circuit systems differ in that the return fluid (often water, or sometimes water mixed with glycol) circulates through the tower structure in a coil, while cooling tower water recirculates only in the tower structure itself (Fig. 19.1). In this case, the return fluid is not exposed directly to the air.

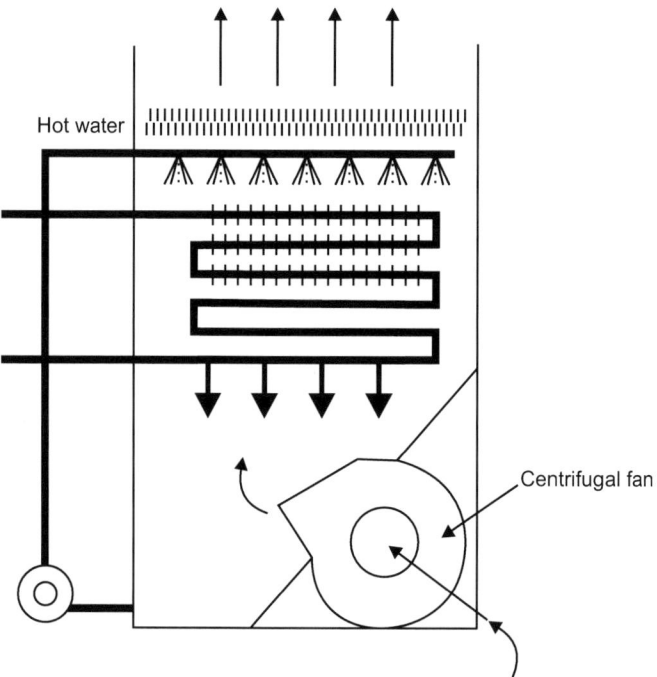

Figure 19.1: Example of a closed circuit cooling tower.

Cooling towers are the primary component used to exhaust heat in open recirculating cooling systems. They are designed to maximise air and water contact to provide as much evaporation as possible. This is accomplished by maximising the surface area of the water as it flows over and down through the tower structure. Figure 19.2 illustrates the different components of a cooling tower structure.

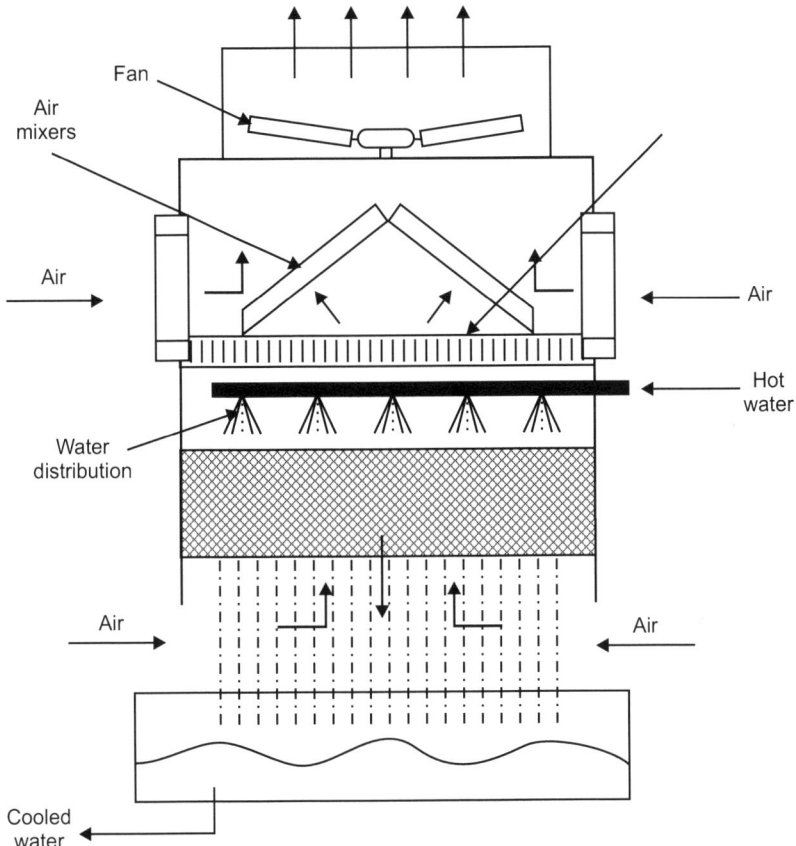

Figure 19.2: Overview of cooling tower structure components.

First, the water is distributed evenly across the top of the cooling tower structure. Tower distribution decks can be a series of spray nozzles oriented up or down (like a landscaping sprinkler system) to uniformly distribute the water over the tower structure. In some cases, the distribution deck may just be a series of holes through which the water falls onto the tower structure. Regardless the distribution deck must uniformly apportion the recirculating water across the tower structure. Broken nozzles or plugged orifices will impede uniform distribution across the tower structure, negatively impacting the overall heat exchange capacity of the system. As the water falls from the distribution deck, the surface area is further expanded in the fill section. Older tower systems may feature splash bars made of plastic, fibreglass or wood that serve to break the falling water into tiny droplets. In recent years, many different forms of labyrinth-like packing or film fill have been incorporated. The closely packed nature of film fill causes the water to travel through this portion of the tower in thin streams, improving thermal efficiency and the evaporation rate, thereby increasing heat rejection.

There are several variations of film fill geometries commercially available and though they do greatly increase the heat rejection rate over splash-type fills, they are also much more susceptible to fouling, scaling and micro-biological growth (these are discussed in greater detail in the system concerns section). Development of any of these problems greatly reduces the cooling efficiency and in severe cases can collapse portions of the tower fill or tower structure. To avoid this, film fill should be inspected routinely to ensure it is clean and free of debris, scale and biological activity.

To minimise losses due to drift and help direct airflow into the tower, louvers and drift eliminators are commonly used. Louvers are most often seen along the sides of the tower structure, while drift eliminators reside in the top section of the tower to capture entrained water droplets that may otherwise leave through the stack. Damaged or incorrectly oriented louvers along with damaged drift eliminators will lead to excessive losses due to drift from the tower structure. Therefore, louvers and drift eliminator sections should be routinely inspected and repaired to ensure optimal water usage.

After the water passes through the fill it cascades down to a collection basin at the base of the tower structure. From the basin the cold water can be pumped back into the system to extract process or comfort cooling needs and begin the cycle all over again.

By design, cooling towers consume large volumes of water through the evaporation process to maintain comfort cooling or process cooling needs, although they use significantly less water than similar capacity once-through cooling systems. Because the evaporative loss is water containing little to no dissolved solids, the water remaining in the cooling tower becomes concentrated with dissolved solids, which can lead to scaling and corrosive conditions. To combat these problems, water with high total dissolved solid content must be drained from the system, via, 'blow down.' The associated losses caused by blow down, evaporation, drift and system leaks must be accounted for by system makeup requirements. Figure 19.3 illustrates these water system flow locations in generic terms.

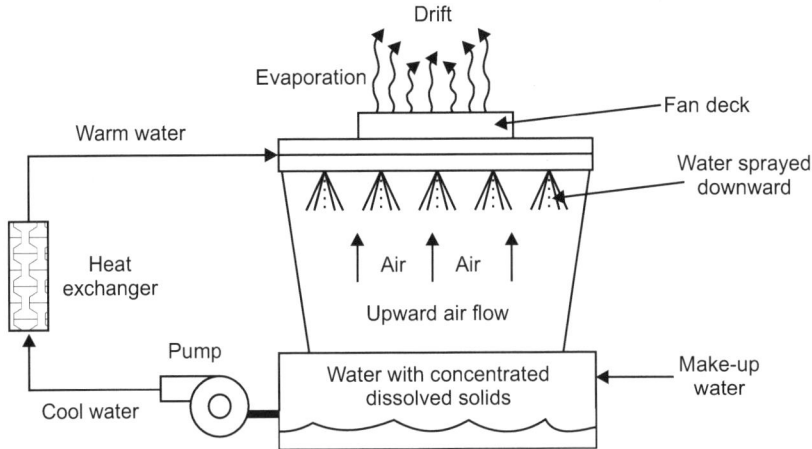

Figure 19.3: Illustration of water flow across a cooling tower.

19.2 Design specifications of cooling towers

Now let us look at where we provide design margins. In sizing pipelines, one has to consider the total flows required and make a provision for design margin of say 10% on flow or the values agreed in the design basis. More often than not while sizing pipelines, the velocities considered are slightly conservative. Thereafter, one needs to take the closest commercially available higher size. If the calculated value is 6.8", one may end up selecting an 8" line. So invariably, this results in a cooling water piping network with varying pressure drops and margins in various sections of the plant. In the course of designing heat exchangers, one may end up with exchangers with pressure drops much lesser than the allowable pressure drop due to allowances for fouling and overdesign requirements of the process. This despite one wanting or having specified a fixed pressure drop across the exchanger.

Another important aspect that affects pressure drops is the equipment and pipeline layout. Generally, all heat exchangers are located in the plant as per the process and layout requirements. Having finalised the layout, one normally calculates the highest pressure drop in the system and accordingly selects the pumps to deliver the required flow and head.

But what happens actually is because of the overdesign in pipelines and exchangers along with the layout requirements, one ends up getting more flow than desired in some exchangers and less flow than desired in others. Although not preferred, one attempts to manage small flow corrections at site by throttling flow using valves. Throttling flows using globe valves is possible for smaller exchangers but cumbersome for larger exchangers and also not recommended. Another method of flow correction is using flow restriction orifices. Doing these orifice calculations at site is difficult (inserting orifice in one flow path changes the flows in other paths). Inserting orifices in big lines/ FRP lines is also troublesome after the lines have been erected. Now software tools like PIPENET®, pipeflow, etc., are available for analysis of networks and using these tools, one can actually create a model of the actual plant with pipe sizes, pipe lengths, bends, elevations, exchangers, etc.

With these tools, it is possible to create a model and analyse the network and optimise the pipe sizes. Spool pieces can be provided in possible problem areas and one can size the orifices and know the flow passing through all the consumers. For some cases, it may even be worthwhile to consider temperature control valves.

This method of analysing and fine tuning the cooling water network can also help in understanding operations during turndown and also provide suitable solutions to save power in the design stage itself. One can thus analyse whether it is feasible to shut off one pump and still meet process requirements or whether providing a variable frequency drive (VFD) for the cooling water pumps is a possible solution. Alternatively, one can evaluate whether one needs the motor to be sized for end of the curve for this operation.

Analysing these flows during turndown is also important, as reducing flow in some exchangers with large overdesign can actually increase fouling due to lower velocities. It is common that velocities in exchangers located at grade are higher than exchangers located at a higher elevation.

Sometimes, one is also constrained to use cooling water on the shell side of exchangers and it is important to ensure that one doesn't end up with low velocities in such exchangers. This enables one to analyse the network and ensure that maldistribution of flow does not result in increased fouling in exchangers due to low velocities. For older plants where one needs to add capacity, creating a model is the right way to optimise and decide on the additional cooling water requirements. Existing pumps can be analysed with regard to their capabilities and one can check for the possibility of using a larger impeller and evaluate the power required from the pump curves.

The requirement for new pumps can be analysed and their capacity specified. It may sometimes be necessary to check the actual flows in the existing plant using a strap-on ultrasonic meter.

19.3   Types of cooling towers

Cooling towers: Cooling towers are classified by the direction of air flow (counter-flow or cross-flow) and the type of draft (mechanical or natural).

Mechanical draft towers: Mechanical draft towers have air forced through the structure by a fan. The air flow can be pushed through by fans located at the base of the tower (referred to as forced draft), or pulled through by fans located at the top of the tower (referred to as induced draft). Induced draft towers tend to be larger than forced draft units.

Natural draft towers: Natural draft towers are designed to move air up through the structure naturally without the use of fans. They use the natural law of differing densities between the ambient air and warm air in the tower. The warm air will rise within the chimney structure because of its lower density drawing cool ambient air in the bottom portion. Often times these towers are very tall to induce adequate air flow and have a unique shape giving them the name 'hyperbolic' towers.

Cross-flow towers: Cross-flow cooling towers are structured so that air flows horizontally across the falling water (Fig. 19.4). This design provides less resistance for the air flow, thereby reducing the fan horsepower required to meet the cooling demand. These towers usually feature gravity-fed water distribution decks that are either open and uncovered, or that are covered to limit algae growth and debris from getting into the distribution deck. Gravity-fed distribution decks have evenly spaced openings that the water drops through to be spread across the tower fill.

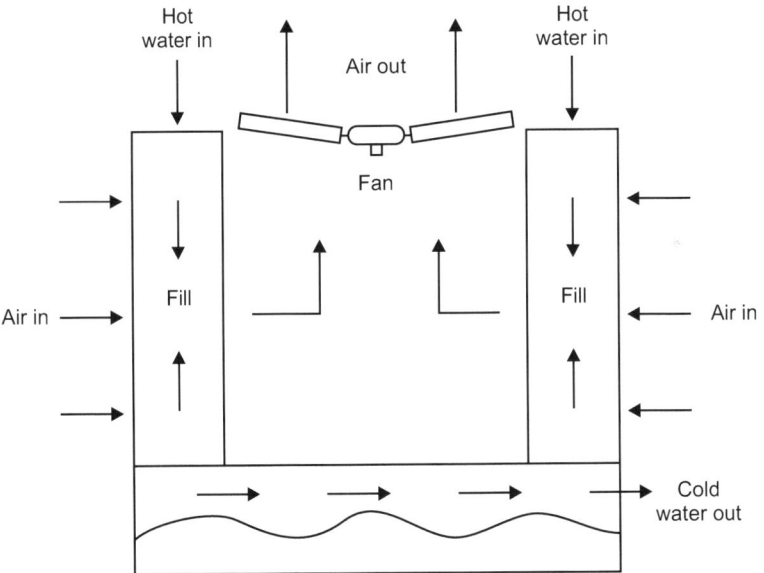

Figure 19.4: Example cross-flow cooling tower.

Counter-flow towers: Counter-flow cooling towers have upward air flow that directly opposes the downward flow of the water providing very good thermal efficiency because the coolest air contacts the coolest water (in the bottom section of the tower structure).

19.4   System calculations of water in cooling tower

To properly operate and maintain a cooling tower, there needs to be a basic understanding of the system water's use. Water use of the cooling tower is the relationship between makeup, evaporation and blow down rates. There are a couple simple mathematical relationships between the blow down rate, evaporation rate, makeup rate and cycles of concentration of a cooling tower that are very useful to understand the principal flow rates. The first relationship illustrates the overall mass balance consideration around a given cooling tower:

1. Makeup = Blow down + Evaporation.

    In this case, the blow down accounts for all system losses including leaks and drift, except for evaporation. The second principal relationship defines cycles of concentration in terms of makeup flow and blow down flow.

2. Cycles of concentration = Makeup ÷ Blow down.

    This equation can be rearranged to either of the following to solve for the makeup rate or blow down rate.

3. Blow down = Makeup ÷ Cycles of concentration.

4. Makeup = Cycles of concentration × Blow down.

    If the evaporation rate and cycles of concentration are known, the blow down rate can then be determined by substituting serial no. 4 into serial no. 1.

5. Cycles of concentration × Blow down = Blow down + Evaporation. Solving for blow down:

6. Blow down = Evaporation ÷ (Cycles of concentration −1).

7. Also, if the blow down rate and cycles of concentration are known, the makeup rate can be determined by solving serial no. 4 and then the evaporation rate can be determined by solving serial no. 6 for evaporation:

8. Evaporation = Blow down × (Cycles of concentration − 1).

19.5   Controlling of cooling tower return temperature and energy saving

Fans, pumps, blowers, cooling towers and many other applications are subject to varying loads. The variation may occur due to various factors, e.g., in cooling tower applications the load variation occurs may be due to utilisation of the installed capacity, variation of process conditions, etc.

Conventionally, control valve throttling, pump discharge bypass recirculation control, etc., are used in order to match the requirements of varying loads in cooling tower outlet temperature control.

In this section it has been shown that apart from conventional mode of controlling varying load can also be significantly reduced by the use of variable frequency drive. The importance of maintaining optimum cooling tower cold water return temperature need not be over emphasised. It is very important for the economic reasons, i.e., from the point of view of desired product yield and quality. Cooling towers are designed to take care of maximum adverse conditions. But normally both of these conditions do not occur simultaneously. Hence, there is lots of scope to improve the performance of operating cooling towers. The design cold water inlet temperature in the secondary circuit of the cooling system, to various systems is approximately between 31–32°C and the maintenance of this design temperature was very important from the point of view of process performance.

However, due to the wide, fluctuation of weather conditions and also due to the partial loading of cooling tower, it had virtually become impossible to maintain optimum cooling tower cold water return temperature, i.e., 28–29°C and operating cold water inlet temperature to various system, i.e., 31–32°C.

19.5.1   Regulation of air loading in cooling tower

Regulation of air loading in cooling tower can be done by the following methods as described below:

1. By changing the length of the fan blade: To reduce the cooling capacity if fan blade length is reduced then its cooling capacity gets fixed to a lower value and it cannot work under varying load. Alternatively, to meet the requirements of variable load at least 2–3 different size blades are to be stocked and blades of operating cooling towers are to be changed as and when load conditions vary. Practically, it is a difficult proposition. Hence, this method is not adopted.

2. Changing the pitch of the fan blade: In order to vary the cooling capacity, according to this method, the pitch of the blades are usually rotated to + or − 3° from the mean. In summer, the blades shall be in, +3° position and in winter −3° position. In winter, −3° position delivers about 80 % of +3° air quantity, the power saving is about 40. However, the limitation in this method is that the regulation of air loading can be done only for two fixed values. Regulation of air loading for a wide range of values is not possible. Hence, this method also was not adopted.

3. Regulating the speed of the fan: Regulating the supply frequency of the power to the fan motor can regulate the speed of the cooling tower fan. By this method, the fan speed and the quantity of air being supplied to the cooling tower can be varied in the entire range from zero speed to maximum fan speed and from zero air supply to maximum air supply as per the requirement of the process. The speed regulation of the fan is possible with the help of advice, viz., variable frequency drive system.

By varying the frequency of the supply power with the help of VDF, the speed of industrial motors (used for running fans, bowlers, pumps, compressors, etc.), can be varied.

This device, VDF, consists of a frequency converter, which varies from frequency and voltage of the supply fed to the AC induction motor. Before buying and installing a VDF for achieving the optimum CTW return temperature, it is advisable to operate the cooling tower at zero fan speed and at full fan speed in order to see the extent the temperature control possible with the installation of a VFD.

19.6 Best management practice: Cooling tower management

Cooling towers regulate temperature by dissipating heat from recirculating water used to cool chillers, air-conditioning equipment or other process equipment. Heat is rejected from the tower primarily through evaporation. Therefore, by design, cooling towers consume significant amounts of water.

The thermal efficiency and longevity of the cooling tower and equipment used to cool depend on the proper management of water recirculated through the tower. Water leaves a cooling tower system in any one of four ways:

1. Evaporation: This is the primary function of the tower and is the method that transfers heat from the cooling tower system to the environment. The quantity of evaporation is not a subject for water efficiency efforts (although improving the energy efficiency of the systems your cooling will reduce the evaporative load on your tower).

2. Drift: A small quantity of water may be carried from the tower as mist or small droplets. Drift loss is small compared to evaporation and blow down and is controlled with baffles and drift eliminators.

3. Blow down or bleed-off: When water evaporates from the tower, dissolved solids (such as calcium, magnesium, chloride and silica) are left behind. As more water evaporates, the concentration of dissolved solids increases. If the concentration gets too high, the solids can cause scale to form within the system or the dissolved solids can lead to corrosion problems. The concentration of dissolved solids is controlled by blow down. Carefully monitoring and controlling the quantity of blow down provides the most significant opportunity to conserve water in cooling tower operations.

4. Basin leaks or overflows: Properly operated towers should not have leaks or overflows. Check float control equipment to ensure the basin level is being maintained properly and check system valves to make sure there are no unaccounted for losses.

The sum of water that is lost from the tower must be replaced by makeup water:

Makeup = Evaporation + Blow down + Drift

A key parameter used to evaluate cooling tower operation is 'cycles of concentration' (sometimes referred to as cycles or concentration ratio). This is calculated as the ratio of the concentration of dissolved solids (or conductivity) in the blow down water compared to the makeup water. Since dissolved solids enter the system in the makeup water and exit the system in the blow down water, the cycles of concentration are also approximately equal to the ratio of volume of makeup to blow down water.

From a water efficiency standpoint, you want to maximise cycles of concentration, which will minimise blow down water quantity and reduce makeup water demand. However, this can only be done within the constraints of your makeup water and cooling tower water chemistry. Dissolved solids increase as cycles of concentration increase, which can cause scale and corrosion problems unless carefully controlled. In addition to carefully controlling blow down, other water efficiency opportunities arise from using alternate sources of makeup water. Water from other equipment within a facility can sometimes be recycled and reused for cooling tower makeup with little or no pretreatment, including the following:

1. Air handler condensate (water that collects when warm, moist air passes over the cooling coils in air handler units). This reuse is particularly appropriate because the condensate has a low mineral content and typically is generated in greatest quantities when cooling tower loads are the highest.
2. Water used in a once through cooling system.
3. Pretreated effluent from other processes, provided that any chemicals used are compatible with the cooling tower system.
4. High-quality municipal wastewater effluent or recycled water (where available).

19.6.1   Operation and maintenance

To maintain water efficiency in operations and maintenance, one should:

1. Calculate and understand your 'cycles of concentration.' Check the ratio of conductivity of blow down and makeup water. Work with your cooling tower water treatment specialist to maximise the cycles of concentration. Many systems operate at two to four cycles of concentration, while six cycles or more may be possible. Increasing cycles from three to six reduces cooling tower makeup water by 20% and cooling tower blow down by 50%.

2. The actual number of cycles you can carry depend on your makeup water quality and cooling tower water treatment regimen. Depending on your makeup water, treatment programmes may include corrosion and scaling inhibitors along with biological fouling inhibitors.

3. Install a conductivity controller to automatically control blow down. Working with your water treatment specialist, determine the maximum cycles of concentration you can safely achieve and the resulting conductivity (typically measured as microSiemens per centimeter, $\mu S/cm$). A conductivity controller can continuously measure the conductivity of the cooling tower water and discharge water only when the conductivity set point is exceeded.

4. Install flow meters on makeup and blow down lines. Check the ratio of makeup flow to blow down flow. Then check the ratio of conductivity of blow down water and the makeup water (you can use a handheld conductivity meter if your tower is not equipped with permanent meters). These ratios should match your target cycles of concentration. If both ratios are not about the same, check the tower for leaks or other unauthorised draw-off. If you are not maintaining target cycles of concentration, check system components including conductivity controller, makeup water fill valve and blow down valve.

5. Read conductivity and flow meters regularly to quickly identify problems. Keep a log of makeup and blow down quantities, conductivity and cycles of concentration. Monitor trends to spot deterioration in performance.

6. Consider using acid treatment such as sulphuric, hydrochloric, or ascorbic acid where appropriate. When added to recirculating water, acid can improve the efficiency of a cooling system by controlling the scale buildup potential from mineral deposits. Acid treatment lowers the pH of the water and is effective in converting a portion of the alkalinity (bicarbonate and carbonate), a primary constituent of scale formation, into more readily soluble forms. Make sure workers are fully trained in the proper handling of acids. Also note that acid overdoses can severely damage a cooling system. The use of a timer or continuous pH monitoring via instrumentation should be employed. Additionally, it is important to add acid at a point where the flow of water promotes rapid mixing and distribution. Be aware that you may have to add a corrosion inhibitor when lowering pH.

7. Select your water treatment vendor with care. Tell vendors that water efficiency is a high priority and ask them to estimate the quantities and costs of treatment chemicals, volumes of blow down water and the expected cycles of concentration ratio. Keep in mind that some vendors

may be reluctant to improve water efficiency because it means the facility will purchase fewer chemicals. In some cases, saving on chemicals can outweigh the savings on water costs. Vendors should be selected based on 'cost to treat 1000 gallons makeup water' and highest 'recommended system water cycle of concentration.' Treatment programmes should include routine checks of cooling system chemistry.

8. Consider measuring the amount of water lost to evaporation. Some water utilities provide a credit to the sewer charges for evaporative losses, measured as the difference between metered makeup water minus metered blow down water.

9. Consider a comprehensive air handler coil maintenance programme. As coils become dirty or fouled, there is increased load on the chilled water system to maintain conditioned air set point temperatures. Increased load on the chilled water system not only has an associated increase in electrical consumption, it also increases the load on the evaporative cooling process, which uses more water.

19.6.2   Retrofit options in cooling towers

The following retrofit options help maintain water efficiency across facilities:

1. Install a sidestream filtration system composed of a rapid sand filter or high-efficiency cartridge filter to cleanse the water. These systems draw water from the sump, filter out sediments and return the filtered water to the tower. This enables the system to operate more efficiently with less water and chemicals. Side-stream filtration is particularly helpful if your system is subject to dusty atmospheric conditions. Side-stream filtration can turn a trouble some system into a more trouble-free system.

2. Install a makeup water or side-stream softening system when hardness (calcium and magnesium) is the limiting factor on cycles of concentration. Water softening removes hardness using an ion exchange resin and can allow you to operate at higher cycles of concentration.

3. Install covers to block sunlight penetration. Reducing the amount of sunlight on tower surfaces can significantly reduce biological growth such as algae.

4. Consider alternative water treatment options, such as ozonation or ionisation, to reduce water and chemical usage. Be careful to consider the life-cycle cost impact of such systems.

5. Install automated chemical feed systems on large cooling tower systems (over 100 T). The automated feed system should control blow down/bleed-off by conductivity and then add chemicals based on makeup water flow. These systems minimise water and chemical use while optimising control against scale, corrosion and biological growth.

19.6.3 Replacement options

The following replacement options will help maintain water efficiency across facilities:

1. Get expert advice to help determine if a cooling tower replacement is appropriate. New cooling tower designs and improved materials can significantly reduce water and energy requirements for cooling. However, since replacing a cooling tower involves significant capital costs, you should investigate every retrofit, operations and maintenance option available and compare costs and benefits to a new tower.

19.7 Energy conservation tips for cooling tower

Following are the brief tips for energy conservation in cooling towers.

1. Control cooling tower fans based on leaving water temperatures.
2. Control to the optimum water temperature as determined from cooling tower and chiller performance data.
3. Use two-speed or variable-speed drives for cooling tower fan control if the fans are few.
4. Stage the cooling tower fans with on-off control if there are many.
5. Turn off unnecessary cooling tower fans when loads are reduced.
6. Cover hot water basins (to minimise algae growth that contributes to fouling).
7. Balance flow to cooling tower hot water basins.
8. Periodically clean plugged cooling tower water distribution nozzles.
9. Install new nozzles to obtain a more-uniform water pattern.
10. Replace splash bars with self-extinguishing PVC cellular-film fill.
11. On old counter flow cooling towers, replace old spray-type nozzles with new square-spray ABS practically-non-clogging nozzles.
12. Replace slat-type drift eliminators with high-efficiency, low-pressure drop, self-extinguishing, PVC cellular units.
13. If possible, follow manufacturer's recommended clearances around cooling towers and relocate or modify structures, signs, fences, dumpsters, etc., that interfere with air intake or exhaust.
14. Optimise cooling tower fan blade angle on a seasonal and/or load basis.
15. Correct excessive and/or uneven fan blade tip clearance and poor fan balance.
16. Use a velocity pressure recovery fan ring.
17. Divert clean air-conditioned building exhaust to the cooling tower during hot weather.

18. Reline leaking cooling tower cold water basins.
19. Check water overflow pipes for proper operating level.
20. Optimise chemical use.
21. Consider side stream water treatment.
22. Restrict flows through large loads to design values.
23. Shut off loads that are not in service.
24. Take blow down water from the return water header.
25. Optimise blow down flow rate.
26. Automate blow down to minimise it.
27. Send blow down to other uses (Remember, the blow down does not have to be removed at the cooling tower. It can be removed anywhere in the piping system.)
28. Implement a cooling tower winterisation plan to minimise ice buildup.
29. Install interlocks to prevent fan operation when there is no water flow.
30. Establish a cooling tower efficiency-maintenance programme. Start with an energy audit and follow-up, then make a cooling tower efficiency maintenance programme a part of your continuous energy management programme.

19.8 Pump energy efficiency for industrial cooling systems

When it comes to industrial cooling systems, a number of plants are running with systems that rely on a fixed-speed centrifugal pump to circulate the cooling medium and a control valve to throttle the flow to a required rate. This type of system is highly inefficient and not only wastes electricity, which has risen dramatically in cost for most industrial users, but it typically requires higher initial capital costs and results in excessive maintenance costs. In a fixed-speed design, the pump is often oversized to ensure that the maximum system requirements are met when the control valve is positioned to accommodate its upper control limit. For most valves, this position is at about 20% closed, which means that the system already has inefficiencies built in because of the required valve restriction.

This problem is compounded when systems are designed for future expansion. The possibility of higher flows requires larger pumps and control valves that typically operate in the middle of their total range, which increases the restriction losses. When the pump is operating, the control valve opens and closes, creating a barrier that the pump tries to push through. While this design creates the desired effect of controlling flow, it also results in excessive wear and tear on pump bearings and valve seats, leading to premature pump failure. In addition, this method of controlling flow causes the pump to consume much higher levels of electricity than necessary.

19.8.1 Intelligent pumps

With most plants producing wider ranges of product than ever before, the need for greater system flexibility has also increased. To operate profitably, it is vital to enhance the efficiency of cooling systems. This goal can be achieved by using intelligent pump systems that vary their own speed to adjust flow, rather than relying on fixed-speed pumps and control valves.

Intelligent pumps, also referred to as E-pumps, use a sensor to electronically vary motor speed to match system demand. The sensor communicates with an integrated frequency drive to match the pump speed precisely to the process demand. These next-generation systems are suitable for designs requiring a single pump, as well as systems requiring multiple pumps. Engineering new intelligent pump systems is simple and many existing systems can be retrofitted easily to include intelligent-pump capabilities. Advantages of intelligent, variable-speed pumps over fixed-speed systems include lower installation costs and reduced maintenance, but the most compelling argument for variable-speed cooling systems is energy savings.

Researchers have found that energy accounts for more than 85% of the life cycle cost of a pump system. Intelligent pumps can cut energy consumption in half, resulting in significant life cycle savings.

While fixed-speed systems, at best, can realise only maximum efficiency part of the time, intelligent pumps can maintain maximum efficiency over their entire operating range.

This capability makes them ideal for cooling applications with varying demands associated with product, volume or seasonal variations. Even the best fixed-speed pumps tend to operate outside of their optimal efficiency points. This drawback isn't the pump's fault; it's just the way the system works. Variable-speed pumps operate at or near their best efficiency point throughout the entire process range, even if the system is designed for future expansion.

The flow restriction caused by the control valve in throttle system designs is replaced with direct control of the system output, so there is no need for a control valve. This design provides even more benefits down the road by reducing the overall maintenance required to keep the system going. Without the valve, the pump doesn't need to work as hard and will last much longer and of course all valve maintenance is eliminated completely.

Many intelligent pump systems have frequency converters integrated into the pump motors to facilitate installation. Some plants prefer to have all converters mounted independently, so panel- and wall-mounted intelligent pump converters are available. Unlike pumps driven by generic variable-frequency drives, intelligent pump converters are preprogrammed to work specifically with pumps, so they do not require extensive knowledge of programming languages. They don't even require advanced knowledge of pumps because they are designed to be installed using simple system set point data. With only a basic understanding of a particular system's requirements, a system can be customised to monitor and adjust to changes in temperature, pressure or flow.

19.8.2 Application: Intelligent pumps and three-way valves

In cooling systems that have fan coil units or similar devices with three-way valves and therefore almost constant system characteristics, adjusting the pump performance based on the differential pressure will be impossible. In such systems, another control parameter must be used. The best choice is to control the circulator pump according to temperature. The return pipe temperature of the system signals changes in the system load. A decrease in the return pipe temperature (i.e., the temperature approaches the supply pipe temperature) indicates that a large amount of the circulated flow is bypassed in the control valves due to a reduced cooling demand in the system. In such situations, the circulated flow in the system should be reduced to lower the power consumption of the pump. A differential temperature control could be used as an alternative to the constant temperature control of the return pipe temperature. The temperature control of the pump offers the same operational advantages as that of differential pressure control.

19.8.3 Application: Controlling flow in cooling towers

In cooling systems that use cooling towers, hot water (typically 80 to 90°F [27 to 32°C]) is circulated from the condenser/heat exchanger to the tower, where the water is cooled by direct means and by evaporation. The tower's cooling capacity depends both on the amount of air and water flowing through the tower.

Air flow usually is controlled with a fan. When the cooling demand is low, the fan often is stopped and cooling is achieved only by evaporation as the water circulates. When the cooling demand increases, the fan restarts to provide additional cooling. In some cases, the fan might be speed-controlled.

The water flow through the cooling tower usually is controlled with a single- or three-way valve and a pump (non-controlled). The valve controls the flow through the cooling tower and thus the cooling power according to the temperature of the return water.

In the case of three-way valves, the pump operates with an almost constant flow, regardless of the load in the tower. However, if two-way valves are used, the pump will be throttled when the cooling demand is reduced. A more energy-efficient solution would be to use an intelligent pump, which would reduce power consumption when the system is only partially loaded and would make the control valve superfluous.

An intelligent pump that has a built-in PID controller can be connected directly to a temperature transmitter, thereby allowing the pump to provide closed loop control of the cooling system.

(*Note:* To ensure a sufficient function of the nozzles in the cooling tower, a minimum flow is required. If the flow is too low, the nozzles cannot atomise the water satisfactorily. Therefore, the minimum pump speed must be set to ensure that this requirement is always fulfilled.)

E-pump, or intelligent pump systems provide greater system flexibility and enable a cooling system to operate at optimum efficiency.

19.9 Nano particles for reducing cooling tower water consumption

19.9.1 Strategic value

Power plant coolants incorporating nanoparticles with phase-change material (Fig. 19.5) cores that melt to absorb heat from steam turbine condensate and solidify as cooling proceeds promise to reduce overall water consumption by as much as 20%. The improved thermal properties offered by these multi-functional nanoparticles also are expected to decrease coolant flow rates by about 15%, helping lower the associated pumping loads and thus parasitic energy losses.

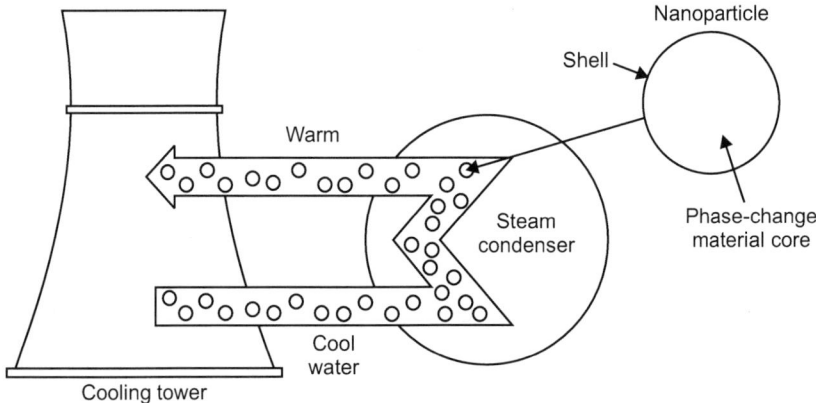

Figure 19.5: Nanoparticles incorporating phase-change materials with wet cooling towers.

Assuming successful development, this innovation will represent a cost-effective and relatively simple retrofit option for increasing water efficiency at operating fossil and nuclear plants with closed-cycle recirculating or hybrid wet-dry cooling systems, as well as a low-cost alternative for reducing the water requirements of new thermoelectric plants. It will help power producers meet water use restrictions at reduced cost and decrease exposure to operational and regulatory risks. It also will increase siting flexibility for new steam-electric generating capacity.

19.9.2   Innovation challenge

U.S. steam-electric plants account for approximately 40% of the nation's total freshwater withdrawals and—due primarily to evaporative and drift losses from wet cooling towers—approximately 3% of overall consumption. High water use rates at plants with closed-cycle wet cooling may not be sustainable in some locations, while thermal discharges from once-through cooling face increasing regulatory scrutiny. Already, some plants operate under water use restrictions or are being required to install water-conserving technologies. Furthermore, the siting of new capacity increasingly is challenged by water supply constraints.

Looking ahead, increasing growth pressures, tightening regulations and changing water balances are expected to represent strong drivers for higher water efficiency in the electric power sector. Current air-cooled steam condensers—the most water-efficient cooling option—have high capital costs, energy penalties and operations and maintenance (O&M) impacts. Advanced technologies for reducing the water requirements of closed-cycle wet and hybrid wet-dry systems represent a critical industry need.

At Argonne National Laboratory (ANL), researchers are developing an innovative nanotechnology-based cooling concept. It involves addition to the coolant stream of nanoparticles with a metallic or ceramic shell encasing a core of phase-change material designed to melt at condenser temperatures and solidify as the heat transfer fluid (HTF) travels through the cooling tower. Adding heat absorption nanoparticles to the coolant is expected to significantly increase its heat transfer coefficient, heating capacity and heat of vaporization. These improved thermophysical properties will allow the same volume of coolant to absorb more heat in the condenser and to dissipate the heat in the cooling tower with lower vaporization. This translates into a reduction in the amount of water required to achieve a given level of cooling, as well as that lost to evaporation and drift.

To support power industry applications, the first priority is to improve understanding of the heat transfer characteristics of these specialised nanoparticles, alone and incorporated within coolants. Next, candidate core/ shell nanoparticles need to be synthesised, evaluated and optimised and improved thermophysical properties demonstrated at the laboratory scale. Practical methods must be developed for adding nanoparticles to power plant cooling circuits to replace losses through drift, evaporation and blow down. Also, impacts on power plant economics and O&M need to assessed and environmental health and safety risks associated with nanoparticle releases need to be evaluated. Ultimately, full-scale field demonstrations will be required.

Carbon footprint in textile industry

20.1   Introduction

The term 'carbon footprint' has become a topic of hot discussion all over the world. Carbon foot print can be described as the extent of damage caused to the environment due to some actions. It is the measure of severity of our activities on the environment, especially on the climate change. Many of the activities in our everyday life produce emissions, through the burning of fossil fuels for electricity, heating, etc. These activities have carbon footprint, producing large amount of greenhouse gases, causing a disastrous effect on the environment.

Greenhouse gases and global warming: Greenhouse gases are produced by human activities, which result in global warming. Carbon dioxide is a major gas that accounts for almost 80% of the emissions. Burning of fossil fuels, oil, natural gas and petrol releases carbon dioxide, methane, nitrous oxide, sulphur hexafluoride, perfluorocarbons, etc., are a few other greenhouse gases originating from industrial processes. These gases accumulate and absorb infrared radiation from the atmosphere, affecting the balance between energy received from the sun and the energy that escapes. The green house gas emission is caused by the production and consumption of fuels, manufactured goods, materials, wood, roads and services. For simplicity of reporting, it is often expressed in terms of the amount of carbon dioxide, or its equivalent of other GHGs emitted. Just as walking on the sand leaves a footprint, burning fuel leaves carbon dioxide in the air, which is called a 'carbon footprint'. Thus the carbon footprint basically relates to the amount of carbon released into the air based on the fuel consumption.

Adverse effects: Due to these emissions, there is a rise in the temperature. During the past 100 years, the earth's temperature has risen considerably. It is estimated that if the current scenario continues, by 2100 global temperature may rise in the range of 1.4–5.8°C. This will result in floods in low coastal areas, unpredictable and extreme weather changes with storms, drought and sudden wild fires. The ecosystem will be disturbed and may put some species to extinction. Vital diseases may spread across the globe.

The carbon footprint is assessed in 2 layers:

1. Primary footprint: Primary footprint monitors carbon emission directly through energy consumption - burning fossil fuels for electricity, heating and transportation, etc.

2. Secondary footprint: Secondary footprint relates to indirect carbon emissions (life cycle of products and sustainability).

20.2　Reducing of carbon foot print in textile industry

The most effective way to decrease a carbon footprint is to either decrease the amount of energy needed for production or to decrease the dependence on carbon emitting fuels.

The textile industry is one of the major consumer of water and fuel (energy required for electric power, steam and transportation). The per capita consumption of textiles is about 20 kg/year and increasing day by day. The world population has reached 7 bn out of which almost 18% is from India.

Thus the energy requirement and consequently the carbon footprint of the textile industry in India is considerably high and at the same time the textile industry in India is expected to grow from an estimated size of US$ 70 bn today to US$ 220 bn by 2020 which would proportionately increase impact on our carbon footprint.

Thus, it is imperative for us to take immediate steps and develop innovative technologies and sustainable solutions that can help reduce the environmental impact. The government is also demanding industries to comply with stricter conditions for environmental protection.

In India also, polyester and cotton constitute more than 80% of textile processing.

The textile industry, according to the U.S. energy information administration, is the 5th largest contributor to CO_2 emissions. Thus the textile industry is huge and is one of the largest sources of greenhouse gases on earth. In 2011, annual global textile production was estimated at 60 bn kg of fabric. The estimated energy and water needed to produce such quantity of fabric is considered to be:

1. 1074 bn kWh of electricity or 132 mn MT of coal.
2. About 6–9 T litres of water.

Thus, the thermal energy required per meter of cloth is 4500–5500 kcal and the electrical energy required per meter of cloth is 0.45–0.55 kWh.

The carbon footprint of the textiles is estimated based on the 'embodied energy' in the fabric, comprising all of the energy used at each step of the process needed to create that fabric. To estimate the embodied energy in any fabric it's necessary to add all the process steps from fibre to finished goods. Based on the fibre used the carbon footprint of various fibres varies a lot.

Further, based on the study done by the stockholm environment institute on behalf of the bioregional development group, the energy used (and therefore the CO_2 emitted) to create 1 T of spun fibre is much higher for synthetics than for cotton:

Fibre	kg CO_2/T of fibre
Polyester	9.52
Cotton-conventional	5.89
Cotton-organic	3.75

For natural fibres, the energy consumption starts at planting and field operations-mechanised irrigation, weed control, pest control and fertilisers (manure vs. synthetic chemicals), harvesting and yields. Synthetic fertiliser use is a major component of conventional agriculture; making one ton of nitrogen fertiliser emits nearly 7 T of CO_2 equivalent greenhouse gases. In case of synthetics, the fibres are made from fossil fuels, where very high amount of energy is consumed in extracting the oil from the ground as well as in the production of the polymers.

The Embodied energy used in production of various fibres

Fibre	Energy in MJ/kg of fibre
Cotton	55
Wool	63
Viscose	100
Polypropylene	115
Polyester	125
Acrylic	175
Nylon	250

Natural fibres, in addition to having a smaller carbon footprint have many additional benefits, being able to be degraded by micro-organisms and composted (improving soil structure); in this way the fixed CO_2 in the fibre will be released and the cycle closed. On the other hand, synthetic fibres do not decompose in landfills they release heavy metals and other additives into soil and groundwater. Recycling requires costly separation, while incineration produces pollutants–in the case of high density polyethylene, 3 T of CO_2 emissions are produced for every 1 T of material burnt.

Substituting organic fibres for conventionally grown fibres considerably helps reduce carbon footprint based on:

1. Elimination of synthetic fertilisers, pesticides and genetically modified organisms (GMOs) which is an improvement in human health and agro-biodiversity
2. Conserves water - making the soil more friable so rainwater is absorbed better–lessening irrigation requirements and erosion.

An additional dimension to consider during processing environmental pollution. Conventional textile processing is highly polluting:

1. Around 2000 chemicals are used in textile processing, many of them known to be harmful to human (and animal) health. Some of these chemicals evaporate while some are dissolved in treatment water which is discharged to our environment.

2. The application of these chemicals uses copious amounts of water. In fact, the textile industry is the largest industrial polluter of fresh water on the planet.

Various ways and methods for reducing the carbon footprint during textile processing have been reported and widely published. Commercially viable products are available in market and being supplied by many organisations. Some of the major areas are discussed below.

20.3 Machinery/equipment related

1. Use of low and ultra low liquor ratio machines–to reduce consumption of water during pre-treatment, dyeing and post dyeing wash-off sequence. Simultaneously reducing the energy required for water heating at various processing steps and effective load on the effluent treatment.

2. Pre-heating of process water by solar panels to reduce consumption of other non renewable energy sources (fossil fuels, wood, husk, etc.).

3. Adequate insulation of dyeing, drying and stenter machines and appropriate heat recovery systems to avoid undesired energy loss.

4. Recycle and reuse of process water and alkali by installing adequate filtration process.

20.3.1 Process related

1. Combined scour and bleach process, combined peroxide neutralising and biosoftening process, one bath one step dyeing of P/C blends, etc., so as to reduce number of textile processing stages and thereby reduce consumption of water and energy.

2. Cold pad batch (CPB) preparation and dyeing for energy conservation.

3. Continuous processing of knits.

4. Pad/dry vs. pad/dry/pad/steam, minimising steam and water consumption during washing processes and minimising number of drying processes.

5. Foam dyeing, finishing and coating.

6. Improving right first time (RFT) and right every time (RET) dyeing performance.

20.3.2   Chemicals and dyes

1. Use of enzymes – biodegradable and non corroding for desizing, scouring, bleach neutralising, biosoftening and post dyeing wash off. Suppliers and formulators of enzymes are offering specialised products for combined processes to reduce number of processing steps.
2. Cationisation of cotton for salt-free dyeing with reactive and direct dyes.
3. High fixation reactive dyeing with reduced salt for exhaustion.
4. Digital ink jet printing.
5. Low temperature curing pigment printing.

Wastewater treatment

1. Use of physical, biological and activated carbon systems.
2. Wastewater treatment sludge used/sold for fuel.

20.4   Sustainability of textiles

The textile industry is a diverse and heterogeneous sector which covers the entire production chain of transforming natural and chemical fibres (such as cotton, wool and oil) into end-user goods, including garments, household goods and industrial textiles.

Textiles are heavily intertwined with environmental, social and governance issues. In the past, efforts of producers and retailers have primarily focused on improving the social aspects of textiles, e.g., establishing fair working conditions, setting social standards, establishing minimum wages, ensuring occupational safety, imposing a ban on child and forced labour, etc.

The precise environmental impact of textiles varies significantly depending on the type of fibre the garment is made from.

However, generally speaking they include:

1. Energy use, greenhouse gas (GHG) emissions, nutrients releases (leading to eutrophication) and ecotoxicity from washing (water heating and detergents) and dying of textiles.
2. Energy use, resource depletion and GHG emissions from processing fossil fuels into synthetic fibres, e.g., polyester or nylon.
3. Significant water use, toxicity from fertiliser, pesticide and herbicide use, energy use and GHG emissions associated with fertiliser generation and irrigation systems related to production of fibre crops, e.g., cotton.
4. Water use, toxicity, hazardous waste and effluent associated with the production stage, including pre-treatment chemicals, dyes and finishes.

All factors along the supply chain have a role to play in reducing the environmental footprint of textile products. First of all producers, because as explained above, considerable impacts might be generated during the fibre production, dying, printing and finishing; but also consumers as considerable environmental impacts occur during the use phase. For example, most of the energy used in the life-cycle of a cotton T-shirt is related to post-purchasing washing and drying at high temperatures. It is also estimated that consumers, in the U.K. throw away as much as 1 million tonnes of textiles every year. Against this background, many voluntary initiatives to reduce the environmental footprint of textiles, especially for cotton and polyester, have been developed or are in the pipeline. The uptake by retailers of the various initiatives in this domaine are high. The 'march' towards more sustainable textiles is well underway. Either as a raw material, as a semi-finished product or as an end product, textiles are assimilated into, or constitute in their own right, a vast range of products used in different domains and for different purposes. This section discusses the most common textiles sold by retail companies: namely clothing and accessories and interior and decoration textiles such as floor coverings, upholstery, curtains, mattresses, household textiles, etc.

Most textiles specific EU legislation addresses the issues of imports from low-wage countries, sets standards for textile names or sets standards for the chemical analysis of textile fibres.

From an environmental perspective, the most relevant pieces of legislation are chemical related: the most important being registration, evaluation, authorisation and restriction of chemical (REACH) substances. For textiles produced in Europe, substances incorporated in the textiles, need to be registered. For imported (outside of the EU) textiles, importers need to notify ECHA if the textiles they import contain SVHC (substances of very high concern) in concentration above 0.1% (w/w) if the total annual volume in all products imported is greater than 1 T. Consumers also have the possibility to ask retailers if products contain SVHC in a concentration above 0.1%.

Unlike REACH and the biocides regulation, the waste framework directive specificially refers to textiles. Besides defining the waste hierarchy, i.e., prevention, preparation for reuse, recycling, energy recovery and disposal, the directive also calls for end of waste specific criteria for textiles to be developed.

Currently, it is the producers and retailers who are mostly driving the improvements in sustainability of textiles and are also working at raising consumer awareness. There is growing attention towards not only social, but also environmental impacts of textiles, especially for specific kind of products such as childrens wear, demand for more environmentally friendlier textiles is continuously increasing. Permanent and quick changes in fashion can be an opportunity for rapid uptake of sustainable garments, but also a barrier since such trends could quickly be replaced by something else. In other areas like interior or underwear innovation cycles are much slower.

20.4.1   Opportunities

1. By improving their environmental and social performances, brands can improve their reputation.
2. Linking business to social and environmental projects enables companies to build a strong connection with consumers by involving them in sustainability initiatives.
3. Technological innovation in production processes, along the supply chain which contribute to improve the environmental footprint of processes and which may save costs, enabling the use of more recycled materials, i.e., end of life polyester can be recycled into new clothes.
4. There are already well established environmental labels that producers can apply for to prove their superior environmental performances (such as the EU ecolabel, blue angel, nordic swan, GOTS).

20.4.2   Barriers

1. Complex and global value chains often with low traceability represent an obstacle for producers and brands who want to improve their production patterns.
2. Socially and environmentally friendlier textiles might result in more expensive finished products.
3. The perception of some consumers that sustainable garments are not stylish or fashionable and that the design and the appearance of eco-clothing is unfashionable and unattractive.
4. An insufficient consumer demand. Producers and retailers who want to promote more environmentally friendlier textiles need to develop the market.
5. The market for recycled garments and fibres is still weak due to insufficient take-back systems and absence of convenient and reputable drop-off locations for unwanted clothing/textiles in many countries, which results in perfectly useable garments sent to landfill or incinerated.
6. Low knowledge level about strategic sustainability among fashion and textile companies and their suppliers and lack of resources to upgrade and integrate new knowledge and new technologies, especially in small and medium sized enterprises.
7. There are many labels on the market which can lead to consumer confusion.

Thus, developing production processes using lower amounts of water, pesticides, insecticides, hazardous chemicals or lower releases of GHG, etc., is as important as the measures adopted by retailers and consumers to select such textiles. However, consumer behaviour in how they care for and dispose of clothing and other textile products is of equal importance, e.g., selecting the appropriate washing temperatures, taking the right steps to significantly extend the lifetimes and encouraging recycling of garments once they have reached their end of life. These important issues are all areas where retailers can have a high degree of influence.

Key challenges

1. Continuing to improve the working and social conditions of workers outside the EU, while offering textiles at an affordable price for EU consumers whose purchasing power is declining.
2. Improving the overall environmental footprint of textiles over their entire life-cycle and supply chain.
3. Changing consumer attitudes of buying as cheap as possible and as many as possible.
4. Providing consumers with relevant information concerning the environmental footprint of the textile products, based on harmonised systems at least at European level.

What can retailers do?

1. Offer and promote more environmentally friendlier textiles.
2. Demand more environmental and social accountability from producers.
3. Communicate to consumers the added value of sustainability and inform them on more environmentally friendly behaviour, e.g., encouraging the most efficient wash cycle programmes, lower temperatures, etc., and how this can help them save money on energy bills and reduce water usage thus lowering overall environmental footprint.
4. Encourage recycling of garments, promoting locally provided clothes banks/bins, etc.
5. For retailers who provide employees with working clothes, revert to more socially and environmentally friendlier textiles.
6. Include sustainability issue in staff training.

What can producers do?

1. Source their suppliers based on their social and environmental performances.
2. Use best practices in technological innovation which contribute to improve the environmental footprint of processes.
3. Substitute hazardous substances with safer substances.

4. Increase information exchange with retailers, provide them with information about the latest innovative solutions that help them address their sustainability challenges/objectives.

5. Support the development of product category rules for textiles according to a methodology at least harmonised at European level and use it as a basis for communicating the environmental performance of their products both in Business to Business (B2B) and Business to Consumer (B2C).

6. Develop and offer more environmentally friendlier textiles.

7. Promote the use of more sustainable fibres like organic cotton, recycled fibres, etc.

8. Engage in research about new fibres and materials with lower environmental impacts compared to natural fibres.

9. Improve care labels on products and together with retailers increase focus on consumer communication to promote responsible care.

10. Encourage the reuse/recycling of old clothes and textiles to produce new clothes, rather than using raw materials, promote remanufacturing and fashion upgrades.

11. Communicate to consumers their sustainability efforts.

12. Demand their suppliers to implement international social standards, e.g., ILO standards.

What can policy makers do?

1. Encourage initiatives, project innovation, etc., and provide incentives for the development and take-up of environmentally friendlier textiles.

2. Lead by example by purchasing environmentally friendlier textiles following the green public procurement (GPP) criteria developed at European level.

3. Support the inclusion of social criteria in the existing EU Ecolabel and take an active role in its on-going revision and GPP criteria developed at European level.

4. Encourage the implementation of ILO norms.

5. Support, implement and/or fund consumer awareness and behaviour change campaigns.

6. Support industry and member states in the development of product category rules on textiles based on a harmonised methodology.

7. Examine the use of economic instruments for promoting sustainable consumption of textiles/clothing.

8. Develop measures for better tackling 'greenwashing', i.e., false sustainability claims.

9 Governments should revisit the approach and effectiveness of policy related to chemical use in the fashion and textile industry, including chemicals used in the fibre or garment production processes, no matter if they take place in the European union or not. In addition, new technologies, such as nanotechnology and GMOs, should be thoroughly investigated to determine whether and to what extent they pose a risk to human health and the environment.

What can we do together?

1. Launch and further promote collaborative initiatives to improve the environmental performance of textiles across the supply chain (sustainable design, fibres and fabrics, maximise reuse/recycling/end-of-life management, sustainable cleaning).

2. Start awareness raising campaigns and sharing of experiences in textile processing regions.

3. Manufacturers (clothing, white goods, detergents, etc.), retailers, consumer groups, etc., should carry out campaigns and inform consumers on issues of common interest related to sustainable fashion consumption and work with designers, celebrities and NGOs to help spread the messages on how to be more environmentally friendly, e.g., reducing the temperatures of the wash cycle, etc.

20.5   Creating new green paradigm

To create new green paradigm the textile and apparel industry needs to adopt 3R concept, i.e., reduce, reuse and recycle.

20.5.1   Reduce

Low carbon foot print processes cut costs by reducing waste of raw materials and energy. Water and energy usage reductions by the textile dyeing and finishing sector can help reduce global carbon dioxide emissions. By saving energy and water, the textile industry cannot only save a lot of money, but also help to slow down climate change. Substituting organic fibres for conventionally grown fibres as it uses less energy, no petrochemical-based fertilisers and pesticides for production, emits fewer GHG and supports organic farming (which has myriad environmental, social and health benefits). Other 'greener' alternatives include organic wool, linen, bamboo, hemp, abaca, soyabean fibre, biopolymers and polyester recycled from used clothing.

Some innovative products with smaller carbon footprints

Some innovative products with smaller carbon footprints are:

1. Polymer fibre, made with agricultural feedstocks, provides a 30% CO_2 reduction while its manufacturing process reduces GHG emissions by 63%, compared to conventional nylon made from petroleum.

2. Polymers fibre products with optimised properties including improved dye ability.

3. Bleaching system that can save up to 40% in energy and water use and reduce cotton loss by 50%.

4. After soaping agent for dyeing can reduce the processing time and water consumption compared to the conventional system.

5. The revolutionary air technology for dyeing requires only one-forth of water and also reduces energy and chemicals consumption.

6. Digital printing, using ink from the dyes, wastes neither fabric nor ink and does not use harmful salts and significantly reduces the environmental footprint.

7. Formaldehyde-free pigment printing system, which ensures 'zero add-on' of formaldehyde during production and needs no further treatment.

8. Colour fast finish, is a one-step-process of textile can reduce the processing time and carbon dioxide emissions.

9. C6-based fluorocarbon finish for stain repellency and release.

10. Innovative machine that applies finishes to fabrics using foam, which conserves water.

11. Industrial enzymes, which are basically proteins, replace harsh chemicals used to remove impurities from the fibre or fabric, which reduces energy costs, water consumption and also improves the feel of the fabric.

20.5.2 Reuse

Effluents of chemically treated textiles are discharged in water. Treatment of wastewater obtained from chemically treated textiles is a must. Use of chrome mordant dyeing and limiting the emission of copper, chromium and nickel into water reduces impurities in dyes and pigments. Using dyeing carriers with high chlorine content should be evaded. During the process of bleaching, alternative agents that are less or not hazardous can be used.

Households currently throw out 1.17 million tonnes of textiles each year, most of it clothing, which could be recycled or reused. It's therefore important, whether as designers, retailers or consumers, that we begin to tackle some of these issues that have been highlighted today.

20.5.3 Recycle

The textile and apparel industry should more utilise recycled fibres. The environmental impact of recycling worn-out polyester or cotton waste into new polyester or cotton fibre respectively, for instance, is significantly lower than making that same fibre a new.

A wide range of innovative, sustainable clothing can be made from recycled textiles. We should take care of the ways to combat 'fast-fashion' and to reduce its negative environmental impact as the issues of textile recycling, cheap clothing or 'throwaway fashion' affects us all.

1. Eco circle environmentally friendly closed-loop recycling system chemically converts used polyester products into new polyester raw materials. The reclaimed polyester is of purity comparable to virgin fibres, but the system reduces energy consumption by 84% and CO_2 emissions by 77%. Recycled polyester products include Ecopet polyester fibre made from recycled polyethylene terephthalate (PET) bottles, Eco circle fibres and recycled polyester fibre recreated from used clothing and uniforms.

2. Rayon, which is produced from wood pulp, seems to be an attractive option, but the manufacturing process still consumes large quantities of energy and creates significant amounts of wood waste. Introduced in the early 1990s, Lyocell is also made from wood fibre (harvested from tree farms). It is biodegradable and recyclable and the production process is more sustainable and includes recovery of most chemicals.

3. Ingeo fibre is the first man-made fibre from 100% annually renewable plant sugars, is supplied into apparel, home textile and increasingly the personal care and hygiene (nonwovens) markets.

20.6　Creating a carbon free environment

Role of trees in offsetting emissions: Trees are the gift of nature to filter our air. They absorb carbon-dioxide and release oxygen. Apart from filtering the air, sustainably managed forests aid multiple environmental and socio-economic functions which will be crucial at the global level in creating a sustainable development. They provide recreational, aesthetic and spiritual benefits.

Treating textile effluents: Effluents of chemically treated textiles are discharged in water. Treatment of wastewater obtained from chemically treated textiles is a must. Use of chrome mordant dyeing and limiting the emission of copper, chromium and nickel into water reduces impurities in dyes and pigments. Using dyeing carriers with high chlorine content should be evaded. During the process of bleaching, alternative agents that are less or not hazardous can be used.

Adopting nature's way of life: Natural fibres are a gift to mankind. Using these fibres makes recycling easy and quicker and is also environmentally friendly. Fibres from linen, bamboo and rice straw are a few to name.

Organic cotton: Using organic cotton is integral, as organic soil scrub the atmosphere to global warming gases by trapping the carbon dioxide and converting the same into soil material. In the global warming arsenal, organic farming is a vital tool.

Environmentalists strongly assert that global warming is a real deal and human activities have been causing it. These are a few measures to curtail the perils of carbon footprints impending in the future. Increasing awareness about the dangers will help people to make educated choices of making changes in their lifestyle to make the carbon footprint smaller.

20.6.1 Development of standards and labels

Global recycle standards

This brand new standard was developed to help verify claims regarding recycled products. The gold level requires products to contain 95–100% recycled material, silver requires 70–95% and Bronze contains a minimum of 30%. The definition of 'recycled' under this standard is based on criteria already laid down by scientific certification systems. In addition, the standard contains environmental processing criteria and raw material specification (water treatment and chemical use is based on GOTS and oeko-tex 100) and social responsibility is incorporated – which ensures workers health and safety and upholds workers rights in accordance with International Labour Organisation (ILO) criteria.

In the U.K., the carbon trust, working with continental clothing, has developed the world's first carbon label for clothing. The new label will provide the carbon footprint of the garment, from raw materials and manufacture to use and disposal.

Carbon footprint label

There exist several third party certifications which we think every conscious consumer of fabric should be aware. We should all know what the certification does and doesn't cover.

20.6.2 Global organic textile standard (GOTS)

Global organic textile standard (GOTS) is a tool for an international understanding of environmentally friendly production systems and social accountability in the textile sector. It covers the production, processing, manufacturing, packaging, labelling, exportation, importation and distribution of all natural fibres. That means, for example: use of certified organic fibres, prohibition of all GMOs and their derivatives and prohibition of a long list of synthetic chemicals. Formaldehyde and aromatic solvents are prohibited; dyestuffs must meet strict requirements (i.e., threshold limits for heavy metals, no Azo colourants or aromatic amines) and PVC cannot be used for packaging. A fabric that is produced to the GOTS standards is more than just the fabric. It's a promise to keep our air and water pure and our soils renewed; it's a fabric, which will not cause harm to you or your descendants. An organic fibre fabric processed to GOTS standards is the most responsible choice possible in terms of stewardship of the earth, preserving health, limiting toxicity the load to humans and animals, reducing one's carbon footprint–and emphasising rudimentary social justice issues such as no child labour.

Cradle to Cradle (C2C)'s minimum requirement for certification is that a product be 67% recyclable or biodegradeable. Oeko-tex, green guard and sustainable materials rating technology (SMaRT) are some other examples of these certifications.

20.6.3   Educate consumers to change attitude

Consumer education about the huge carbon footprint mainstream textiles have to help inspire consumers to change their habits is a must. It also assists in changing consumer attitudes. Their inclination towards 'organic fabrics' not simply fabric made from organic fibres, eco-friendly fibres, not cotton or synthetics; minimising purchase of fabrics that are blends of natural and synthetic fibres (i.e., cotton and polyester), or blends of two or more different synthetic fibres (polyester and acrylic), because there is no hope of recycling these fabrics right now. Search for a fabric or product that is certified by any third party, independent textile certification agency - GOTS, SMaRT, C2C, etc., paying attention to the carbon footprint of the fabrics they buy. Keeping themselves educated on the progress of the eco-textile community–are few of the steps that will truly reduce carbon footprint of textile and apparel industry. To maintain and grow their customer base of this new generation of environmental and ethically aware consumers, retailers in particular are pushing sustainability requirements back down their global supply chains.

20.6.4   Low-carbon manufacturing programmes

Low-carbon manufacturing programmes and carbon accounting in factories, carbon footprint calculation projects, benchmarking energy consumption across the textile and apparel supply chain are few of the strategic measures required to reduce carbon footprint of textile and apparel industry in India.

Thus, the global textile industry has taken several strides towards reducing its carbon footprint and meeting the challenges of building a more sustainable future. At the same time there is a growing awareness of environmental issues among consumers who are now increasingly insisting on textile products complying with environmental standards. These complementary trends will hopefully continue to drive the industry toward offering the consumer products that are not only red, blue, white, etc., but also green.

Beyond fibre production, the dyeing and finishing sector is the largest energy and water consumer in the whole textile chain and has the highest potential for energy and water savings and efficiency improvements. Action is needed, but the industry cannot do it alone. National and multinational governments should support the industry with incentive plans to change old technology with modern equipment.

20.7   LCA, carbon footprint and ecological footprint

In these times of climate change concern, individuals and organisations alike are eager for measurable criteria to compare the impact of products and services on global warming.

Life cycle assessment: Life Cycle Assessment (LCA) is the broadest indicator and an internationally standardised method (ISO 14040 and ISO 14044). It not only evaluates the impact on climate change, but also other impact categories such as acidification potential, eutrophication potential, ozone depletion potential and ground level ozone creation. For each of these impact categories, the product or system is evaluated over its complete life span, from the extraction of raw material and manufacturing, to the use of the product by final consumers and end-of-life processes like recycling, energy recovery and ultimate waste disposal.

The ISO standards provide robust and practice-proven requirements for performing transparent LCA calculations. Moreover, one can make use of extensive databases containing life cycle profiles of many goods and services, as well as many of the underlying materials, energy resources, transport systems, etc. Nevertheless, LCA calculations remain very complex and should therefore be applied only by professionals and preferably to a specific unit or application, such as a washing machine or a car tyre.

Carbon footprint: Today, the term 'carbon footprint' is often used as shorthand for the amount of carbon (usually in tonnes) being emitted by an activity or organisation. A carbon footprint, also called carbon profile, is an LCA with the analysis limited to emissions that have an effect on climate change (carbon dioxide, methane, etc.). This limitation makes it easier to apply the calculation to integrated systems.

Ecological footprint: The ecological footprint is a measure of human demand on the earth's ecosystems. It essentially measures the supply and demand of goods and services for an entire planet by assuming that the whole planetary population follows a specific lifestyle of a known person/group of people. The estimate for the ecological footprint begins with the calculation of the land, water/sea needed to support the particular food, shelter, nobility and goods and services needs of a person in a particular region. This estimation changes with the area that person lives.

This is due to the fact that ecosystems vary in their ability to produce useful biological materials and to absorb CO_2, which is called the biocapacity. The results are given in the number of planet earths it would take to support humanity, if everybody follows estimated lifestyle.

20.7.1 Difference between ecological footprint and carbon footprint

The ecological footprint and the carbon footprint are both matrices developed to measure the impact of routine human activity on the environment. Yet they differ in their scope, expression of impact values and the perspective of the calculation. The carbon footprint takes into account only the activities related with green house gas emission. Those are direct methods such as fossil fuel burning and indirect methods such as consumption of electricity. However, the ecological footprint describes all the activities a person is involved in and the resources utilised as well as the wastage generated through the said activity. The carbon foot print gives the raw amount of carbon emission in tonnes per year as an outcome. But the ecological footprint gives values of the land and water area that is needed to replace the resources consumed. Furthermore, the carbon footprint aims to reduce the impact on the environment by reducing global warming and evading catastrophes such as climatic change. But the ecological footprint takes all problems of the environment as a whole and aims for a sustainable development.

The carbon foot print represents the most rapidly growing and most destructive portion of the ecological footprint. Reducing the carbon footprint is the foremost step in reducing the over run consumption of resources. But to get an overall idea of the true impact, which addresses issues such as over fishing, over grazing and deforestation, the ecological foot print is required. Most importantly, statutory bodies should use both these calculators to manage their resources and secure their future.

Role of nanotechnology in energy conservation

21.1   Introduction

Nanoscience and nanotechnology are the study and application of extremely small things and can be used across all the other scientific fields, such as–chemistry, biology, physics, materials science and engineering.

Nanotechnology overcomes the limitation of applying conventional methods to impart certain properties of textile materials. There is no doubt that in the next few years, nanotechnology will penetrate into every area of the textile industry. Nanotextiles are nanoscale fibrous materials that can be functionalised with a vast array of novel properties, including antibiotic activity, self-cleaning and the ability to increase reaction rates by providing large surface areas to potential reactants. In recent years was demonstrated that nanotechnology can be used to enhance textile attributes, such as fabric softness, durability and breathability, water repellency, fire retardancy, antimicrobial properties in fibres, yarns and fabrics. The development of smart nanotextiles has the potential to revolutionise the production of fibres, fabrics or Nonwovens and functionality of our clothing and all types of textile products and applications. Nanotechnology is considered one of the most promising technologies for the 21st century. Today is said that if the IT is the wave of the present, the nano-technology is the wave of the future.

Today there are many who think that the next industrial revolution is right around the corner-because of nanotechnology. They think that nanotechnology will radically transform the world and the people, of the early 21st century. It has the capacity to change the nature of almost every human made object.

21.2   Nanotechnology in the textile industry

Nanotechnology has been discovered by the textile industry-in fact, a new area has developed in the area of textile finishing called 'nanofinishing'. Making fabric with nano-sized particles creates many desirable properties in the fabrics without a significant increase in weight, thickness or stiffness, as was the case with previously used techniques. Nanofinishing techniques include: UV blocking, anti-microbial, bacterial and fungal, flame retardant, wrinkle resistant, antistatic, insect and/or water repellant and self-cleaning properties.

Finishing of fabrics made of natural and synthetic fibres to achieve desirable hand, surface texture, colour and other special aesthetic and functional properties, has been a primary focus in textile manufacturing. In the last decade, the advent of nanotechnology has spurred significant developments and innovations in this field of textile technology. Fabric finishing has taken new routes and demonstrated a great potential for significant improvements by applications of nanotechnology.

There are many ways in which the surface properties of a fabric can be manipulated and enhanced, by implementing appropriate surface finishing, coating and/or altering techniques, using nanotechnology. Today, the main applications of nanotechnology in textiles refer to: Nanofinishing in textiles, nano chemicals for textiles, nanocoating for textile materials, nano/smart silver for textile. In recent years was demonstrated that nanotechnology can be used to enhance textile attributes, such as fabric softness, durability and breathability, water repellency, fire retardancy, antimicrobial properties in fibres, yarns and fabrics.

One of the most common ways to use nanotechnology in the textile industry is to create stain and water resistance. To do this, the fabrics are embedded with billions of tiny fibres, called 'nanowhiskers' (think of the fuzz on a peach), which are waterproof and increase the density of the fabric. The Nanowhiskers can repel stains because they form a cushion of air around each fibre.

Nanotechnology can also be used in the opposite manner to increase the ability of textiles, particularly synthetics, to absorb dyes. Until now most polypropylenes have resisted dyeing, so they were deemed unsuitable for consumer goods like clothing, table cloths, or floor and window coverings. A new technique being developed is to add nanosized particles of dye friendly clay to raw polypropylene stock before it is extruded into fibres. The resultant composite material can absorb dyes without weakening the fabric.

Nanotex (U.S. based company) is a leading fabric innovation company which provides nanotechnology-based textile enhancements to the apparel, home and commercial/residential interiors markets. For example, its product, Aquapel, is the next generation in water repellent, eco-friendly performance, providing advanced protection against rain, sleet, snow and spills. Using a proprietary hydrocarbon technology, aquapel modifies fabric at the molecular level by permanently attaching hydrophobic 'whiskers' to individual fibres, without altering the fabric's natural breathability or feel. Plus, aquapel is fluorocarbon free and problem formulation and option assessment (PFOA) free, making it the right choice for you and the earth.

The other main use of nanoparticles in textiles is that of using silver nanoparticles for antimicrobial, antibacterial effects, thereby eliminating odours in fabrics. Nanoparticles of silver are the most widely used form of nano-technology in use today, says Todd Kuiken.

The silver is made smarter through nanotechnology:

1. Lasts the expected life of the product.
2. Uses the natural antimicrobial action of silver in controlling the growth of odour-causing bacteria, fungus and mold.
3. Is easily integrated into natural and synthetic fibres, foams, plastics and coatings.
4. Has been thoroughly tested and is eco-friendly.
5. Meets regulatory requirements.
6. Has a track record with products in the health care, textile and industrial markets.

The future for textile applications using nanotechnology is exploding due to various end uses like protective textiles for soldiers, medical textiles and smart textiles.

For example, consider the T-shirt. Research is being done that will use nanotechnology enhanced fabric so the T-shirt can monitor your heart rate and breathing, analyse your sweat and even cool you off on a hot summer's day. What about a pillow that monitors your brain waves, or a solar powered dress that can charge your ipod or MP4 player?

Nanotechnology has the potential to being revolution in the field of technical textiles for the benefit of humanity.

21.3   Quality label for nanotechnology

The Hohenstein Institutes, an accredited test laboratory and research institute, which was founded in Bönnigheim (Germany) in 1946, was launched in October 2005 its quality label for nanotechnology, a litmus test as to whether product is nano or not. The certification of the textiles is based on their adherence to a strict definition of nanotechnology which can be applied to the textile sector, developed in conjunction with nanomat, a Germany-based nanomaterials network. 'Nanotechnology refers to the systematically arranged functional structures which consist of particles with size-dependent properties'.

The program and quality label was instituted to help retailers and other textile and users determine if a textile product really incorporates nanotechnology or whether the name, as applied to a particular product, is merely an advertising message. The label offers retailers and consumers guidance in the maze of confusing advertising messages and forms the basis for reliable product comparison.

Testing of nanotechnology includes:

1. Determination of the type of nanotechnological finishing.

2. Visual inspection of nanotechnological finishing using a scanning electron microscope.
3. Quantification of the effect of the finishing (e.g., dirt repellence by measurements of contact angle on characteristic fluids, antimicrobial effects of Nano-Ag, UV protection of Nano-Ti/Nano-ZnO).
4. Determination of mechanical suitability for use.
5. Laundering permanence.
6. Determination of breathability.
7. Determination of biocompatibility.

The testing program is tailored to the textile material and its areas of application. Testing is carried out on new textiles and after simulated conditions of use. The requirements defined for the award of the label are product-specific. For example, for a pair of trousers with a soil-repellent finish, the breathability must not be significantly affected and the skin compatibility must be proven by tests for tissue compatibility. The resistance of the nano-finish to the effects of wear (abrasion resistance) and care are also tested. For care treatments, the stated function is guaranteed for a defined minimum number of washing and drying cycles. The additional parameters are also stated and explained on the hohenstein quality label.

21.4   Application of nanotechnology in textile industry

Due to new technologies, the textile industry has been expanding to many new areas in the past decade, all belonging to the category of technical textiles. Textiles or textile-based composites are expected to replace many metallic and plastic materials used in for example the automobile industry, construction sector, machinery and machine tools industry, electronics and to a lesser extent, in wood, leather and other natural materials in furniture, sport goods and many other smaller application areas.

The main fields of application for textiles are:
1. Agriculture and forestry.
2. Healthcare.
3. Building and construction.
4. Home and household.
5. Clothing.
6. Industry and machinery.
7. Defence.
8. Mobility and transport.
9. Electronics.

10. Packaging.
11. Environmental protection.
12. Sport and leisure.
13. Geotextiles and civil engineering materials.

21.4.1   How nanotechnology works in textiles?

The nanotechnology innovations in the textile industry include both the development of new materials and the improvement of existing materials.

The use of nanotechnology allows textiles to become multifunctional. The so-called 'plasma' technology for instance, is being used to modify the top nanometre layers of textiles, making them antibacterial, water repellent and able to kill fungus at the same time.

Three types of nanotechnology in textiles can be distinguished:

1. Nanotechnology in fibres and yarns (fabrics).
2. Nanotechnology in coatings (textile finishing).
3. E-textiles.

Nanotechnology in fibres and yarns

Nanotechnology creates many new possibilities in fabrics. Nanofibres make it possible to create new blended yarns and fabrics, enhance or alter the properties of textiles, produce synthetic fibres with properties of natural fibres, etc. The difference with normal fibres is that nanofibres have larger surface areas, that can be used to react with the environment.

Molecular (nano) layers and a smaller pore size in the fabrics lead to, for example, self-cleaning and entrapment possibilities.

A different type of nanofibres are nanostructured composite fibres. These fibres contain nanosize fillers such as clay or metal nanoparticles, graphite nanofibres or carbon nanotubes. Another possibility in composite fibres are nano thin coatings around each fibre. The nanosize fillers and coatings are used to increase the mechanical strength and improve the physical properties such as conductivity or antistatic behaviour. Another interesting development are the so-called controlled release polymers. They activate the release of for example antifungals, fragrances or medical growth aids to the fabric. The triggered release systems can be made responsive to stimuli such as changes in temperature, humidity and oxygen levels.

Nanotechnology in coatings

The main advantage of nanotechnology in coatings is that it is no longer visible or perceptible on the product and can still be made more controllable and more thorough. The separate molecules or nanoparticles of the finishes can be brought individually to specific, designated places on textile materials, which wasn't possible on such scale before. This means that much less material is needed to create the same effect. Plus, nano-finishes can no longer be washed off, in contrary to previous textile finishes.

Nanoparticles such as metal oxides and ceramics are used in textile finishing to change surface properties and to add functions to the textile. Because of their larger surface area nanoparticles have a higher efficiency compared to larger size particles. Besides, nanosize particles are transparent and do not blur colour and brightness of the textile substrates.

Nanotechnologies for coatings include: self-healing composites, wear and corrosion resistant coatings, cut resistant material and wrinkle resistant fabric.

E- textiles

E (electronic)-textiles are fabrics made from yarns that carry electronic components. Nanotechnology encourages the development of E-textiles in allowing electronic devices to become smaller and more powerful and creating new electrically conductive fibres and textiles. Nanotechnology treated fibres and fabrics have formed an entire range of smart textiles which can be used in numerous applications, like practical sportswear, medical and safety wear and fashion clothing.

E-textile products include: jackets with integrated audio control systems, heart sensing tops and bras, fabrics containing Light Emitting Diodes (LEDs) that display moving images on textiles, bags containing solar cells for charging mobile phones or iPods, etc.

Nanotechnology in textile finishing

Nanoscale emulsification, through which finishes can be applied to textile material in a more thorough, even and precise manner provide an unprecedented level of textile performance regarding stain-resistant, hydrophilic, anti-static, wrinkle resistant and shrinkproof properties.

Nanosize metal oxide and ceramic particles have a larger surface area and hence higher efficiency than larger size particles, are transparent and do not blur the colour and brightness of the textile substrates. Fabric treated with nanoparticles TiO_2 and MgO replaces fabrics with active carbon, previously used as chemical and biological protective materials. The photocatalytic activity of TiO_2 and MgO nanoparticles can break down harmful chemicals and biological agents. Finishing with nanoparticles can convert fabrics into sensor-based materials. If nanocrystalline piezoceramic particles are incorporated into fabrics, the finished fabric can convert exerted mechanical forces into electrical signals enabling the monitoring of bodily functions such as heart rhythm and pulse if they are worn next to skin.

21.4.2 Opportunities in the textile industry

Nanotechnology can enrich the textile industry in several ways. First of all and most obvious, it creates many new possibilities and improved functions in textiles. New materials can be developed and new properties can be added to existing materials by the use of special-treated fibres, coatings and e-textiles.

This will not only benefit the textile industry itself, but many other industries that make use of some form of textiles. Through this, the textile industry will be able to expand to new markets, as is currently already happening.

Nanotechnology also allows the textile industry to better anticipate to changing consumer needs. More and more, consumers are demanding functions in clothing that go beyond appearance: the quality of products has become more important. Consumers now want the product to smell pleasant, stay fresh, feel comfortable, keep clean and to be simple to care for. Also, information and communication technologies have become very important in daily life.

Nanotechnology offers manufacturers the possibility to combine these changed consumer need with (existing) textile products like sportswear, home textiles and clothing. But nanotechnology is not just a promising new way of working with textiles, in many cases the step towards nanotechnology is part of the strategy that textile companies hope keeps them surviving in a tough textiles market. Especially in the EU textile production, nanotechnology plays an important role because of this.

Increased competition, particularly from Asia, combined with the abolition of all import quotas for textiles and clothing in the EU, US., Canada and Norway in 2008, has forced the industry to streamline and modify. It has become clear that manufacturing traditional products may no longer be enough to maintain a profitable business: the industry has to move in the direction of more innovative, high quality products in order to distinguish and compete.

Much of the focus in the EU is on technical textiles with high added value, in order to develop new markets in construction, protective clothing and medical uses for example. By finding a niche, textile companies will not have to sell lower priced goods that struggle to compete with cheaper textiles from low cost countries.

Thus, nanotechnology has great potential in the textile industry. It will enable growth for the technical textiles market and provide a stronger competitive position by creating innovative, high performance products. Through this, companies in Europe and the U.S. can distinguish themselves from the cheaper producing countries, particularly in Asia.

Nanotechnology creates many possibilities in fibres, coatings and e-textiles and is currently mostly present in non-cost sensitive fields like professional sports, military/protective textiles and medical applications, in which performance is more important than cost reduction.

21.5   Nanotechnology in energy conservation

Nanotechnology provide the potential to enhance energy efficiency across all branches of industry and to economically leverage renewable energy production through new technological solutions and optimised production technologies. Nanotechnology innovations could impact each part of the value-added chain in the energy sector which are discussed below.

21.5.1   Energy sources

Nanotechnology provide essential improvement potentials for the development of both conventional energy sources (fossil and nuclear fuels) and renewable energy sources like geothermal energy, sun, wind, water, tides or biomass. Nano-coated, wear resistant drill probes, for example, allow the optimisation of lifespan and efficiency of systems for the development of oil and natural gas deposits or geothermal energy and thus the saving of costs. Further examples are high-duty nanomaterials for lighter and more rugged rotor blades of wind and tide-power plants as well as wear and corrosion protection layers for mechanically stressed components (bearings, gear boxes, etc.).

Nanotechnologies will play a decisive role in particular in the intensified use of solar energy through photovoltaic systems. In case of conventional crystalline silicon solar cells, for instance, increases in efficiency are achievable by antireflection layers for higher light yield.

First and foremost, however, it will be the further development of alternative cell types, such as thin-layer solar cells (among others of silicon or other material systems like copper/irradium/selenium), dye solar cells or polymer solar cells, which will predominantly profit from nanotechnologies. Polymer solar cells are said to have high potential especially regarding the supply of portable electronic devices, due to the reasonably-priced materials and production methods as well as the flexible design. Medium-term development targets are an efficiency of approximately 10% and a lifespan of several years. Here, for example, nanotechnologies could contribute to the optimisation of the layer design and the morphology of organic semi-conductor mixtures in component structures. In the long run, the utilisation of nanostructures, like quantum dots and wires, could allow for solar cell efficiencies of over 60%.

21.5.2   Energy conversion

The conversion of primary energy sources into electricity, heat and kinetic energy requires utmost efficiency. Efficiency increases, especially in fossil-fired gas and steam power plants, could help avoid considerable amounts of carbon dioxide emissions.

Higher power plant efficiencies, however, require higher operating temperatures and thus heat-resistant turbine materials. Improvements are possible, for example, through nano-scale heat and corrosion protection layers for turbine blades in power plants or aircraft engines to enhance the efficiency through increased operating temperatures or the application of lightweight construction materials (e.g., titanium aluminides).

Nano-optimised membranes can extend the scope of possibilities for separation and climate-neutral storage of carbon dioxide for power generation in coal-fired power plants, in order to render this important method of power generation environmentally friendlier in the long run. The energy yield from the conversion of chemical energy through fuel cells can be stepped up by nano-structured electrodes, catalysts and membranes, which results in economic application possibilities in automobiles, buildings and the operation of mobile electronics.

Thermoelectric energy conversion seems to be comparably promising. Nano-structured semiconductors with optimised boundary layer design contributes to increase in efficiency that could pave the way for a broad application in the utilisation of waste heat, for example in automobiles, or even of human body heat for portable electronics in textiles.

21.5.3   Energy distribution

Regarding the reduction of energy losses in current transmission, hope exists that the extraordinary electric conductivity of nanomaterials like carbon nano-tubes can be utilised for application in electric cables and power lines. Furthermore, there are nanotechnological approaches for the optimisation of superconductive materials for lossless current conduction. In the long run, options are given for wireless energy transport, e.g., through laser, microwaves or electromagnetic resonance. Future power distribution will require power systems providing dynamic load and failure management, demand-driven energy supply with flexible price mechanisms as well as the possibility of feeding through a number of decentralised renewable energy sources.

Nanotechnologies could contribute decisively to the realisation of this vision, inter alia, through nano-sensory devices and power-electronical components able to cope with the extremely complex control and monitoring of such grids.

21.5.4   Energy storage

The utilisation of nanotechnologies for the enhancement of electrical energy stores like batteries and super-capacitors turns out to be downright promising. Due to the high cell voltage and the outstanding energy and power density, the lithium-ion technology is regarded as the most promising variant of electrical energy storage. Nanotechnologies can improve capacity and safety of lithium-ion batteries decisively, as for example through new ceramic, heat-resistant and still flexible separators and high-performance electrode materials.

The company Evonik of U.S. pushes the commercialisation of such systems for the application in hybrid and electric vehicles as well as for stationary energy storage.

In the long run, even hydrogen seems to be a promising energy storage for environmentally-friendly energy supply. Apart from necessary nanostructure adjustments, the efficient storage of hydrogen is regarded as one of the critical factors of success on the way to a possible hydrogen management.

Current materials for chemical hydrogen storage do not meet the demands of the automotive industry, which requires a hydrogen-storage capacity of up to ten weight per cent. Various nanomaterials, inter alia based on nanoporous metal-organic compounds, provide development potentials, which seem to be economically realisable at least with regard to the operation of fuel cells in portable electronic devices. Another important field is thermal energy storage. The energy demand in buildings, for example, may be significantly reduced by using phase change materials such as latent heat stores. Interesting, from an economic point of view, are also adsorption stores based on nanoporous materials like zeolites, which could be applied as heat stores in district heating grids or in industry. The adsorption of water in zeolite allows the reversible storage and release of heat.

21.5.5 Energy usage

To achieve sustainable energy supply and parallel to the optimised development of available energy sources, it is necessary to improve the efficiency of energy use and to avoid unnecessary energy consumption. This applies to all branches of industry and private households. Nanotechnologies provide a multitude of approaches to energy saving.

Examples are the reduction of fuel consumption in automobiles through light-weight construction materials on the basis of nanocomposites, the optimisation in fuel combustion through wear-resistant, lighter engine components and nanoparticular fuel additives or even nanoparticles for optimised tyres with low rolling resistance.

Considerable energy savings are realisable through tribological layers for mechanical components in plants and machines. Building technology also provides great potentials for energy savings, which could be tapped, for example, by nanoporous thermal insulation material suitably applicable in the energetic rehabilitation of old buildings.

In general, the control of light and heat flux by nanotechnological components, as for example switchable glasses, is a promising approach to reducing energy consumption in buildings.

21.6   Nanotechnology helps solve the world's energy problems

The aim is to explain how nanotechnology can help address present and future sustainable energy needs. The main fields of sustainable energy policies and research: renewables, conventional energy, more energy efficiency in industrial production and energy saving.

The relevant technologies and applications include: solar cells, hydrogen and fuel cells, batteries, improvement of light bulbs, fossil fuel, etc., with nanostructured materials and nanopowders, isolation materials, membranes and catalysts, etc.

21.6.1   Solar photovoltaics

Solar Photovoltaic (PV) electricity production is the most obvious technology where nanostructured materials and nanotechnology are contributing to technology development. Solar PV is already competitive in electricity production for homes or villages in remote areas without a connection to the electricity grid. Governments in the US, Europe and Japan are subsidising both technology development and installation of PV modules on roofs and integrated in new buildings for private homes, companies, or even churches (in Germany). The dominant technologies are at the moment mono or multicrystalline silicon. The solar cells are produced by sawing 0.2–0.3 mm thin wafers from lumps of silicon. The problem is that this uses a lot of expensive material, about half of which gets wasted in the sawing process.

Thin film nanostructured alternatives which are currently on the market use an active layer of microns thickness, deposited on a cheap substrate such as glass. These alternatives include amorphous silicon, which is best known from its use in pocket calculators, but is also used in solar panels, on the market for about 15 years. Amorphous silicon is cheaper than crystalline silicon, because it uses 300x less active material. The efficiency is much lower, less than 10% compared to 15%.

Two other available thin film alternatives which entered the market in 2001 are Copper Indium diSelenide (CIS) and Cadmium Telluride CdTe. The market chances of the CdTe technology may be diminished because of environmental concerns. Cadmium is a toxic material. Metallic III-V high performance cells are mostly used in space applications, but also in concentrator cells. Concentrator cells consist of a relatively expensive efficient solar cell and a device which funnels the incoming sunlight from a wider area to the cell. In the lab, efficiencies up to 40% have been measured. But real world manufacturing never achieves the same high efficiencies. One problem with thin film solar PV based on nanotechnology is that energy conversion is even less efficient than in crystalline silicon. According to a spokesperson from BP Solar, the main bottleneck in thin film PV manufacturing is that nobody can produce large enough areas of the thin films on an industrial scale.

Longer term alternatives include the organic Grätzel cell, first invented in 1991 by prof. Michael Grätzel. The principle is also the basis for other research on solid state variants. Prof. Joop Schoonman at the Technical University of Delft, Netherlands aims to replace the liquid electrolyte with a conducting polymer or inorganic material such as FeS, CuS, CuInS.

21.6.2 Grätzel cells

The organic Grätzel solar cell consists of a 10 μm thin layer of Titanium Dioxide TiO_2 particles, which are 20 nm in diameter. Organic dye molecules are adsorbed in the pores between the TiO_2 particles, surrounded by an electrolyte fluid. The cell is completed by two transparent electricity conducting electrodes and a catalyst. The efficiency of Grätzel cells is much lower than of commercial crystalline silicon (around 7–8% instead of around 15%). Therefore they are not competitive in the main market for Solar PV. The EU Nanomax project aims to improve this performance to 15%. Prof. Wim Sinke of ECN in the Netherlands coordinates it. Some start-up companies are already producing Grätzel cells for niche markets.

The company Greatcell www.greatcell.com (now part of Leclanché, a battery producing company in Switzerland) has developed the technology further and now offers its first product. This solar powered clock can work indoors without a battery. It can work indoor, because the organic dye sensitive solar cells can convert low light intensities in electricity.

In Australia, the Sustainable Energy Development Authority (SEDA) is investing US$368,000 in a project to integrate Grätzel organic dye solar cells into the walls of the CSIRO Energy Centre in Newcastle. The start up Sustainable Technologies International Ltd. (STI) will deliver the 200 m^2 PV panels. Nanotechnology is not really difficult, as this example shows: Even a child can make organic solar cells including nanostructured material.

The company Mansolar in the Netherlands manufactures and sells educational kits for school children to make their own organic solar cell, using blackcurrant juice or hibiscus tea as the dye. The company started in 2000 as a spin-off from the Energy Centre Netherlands (ECNs) in Petten.

21.6.3   Hydrogen

There is a lot of discussion at the moment about the Hydrogen Economy, where hydrogen will be the dominant fuel, converted into electricity in fuel cells, leaving only water as waste product. The hydrogen is not freely available in nature in large quantities, so it must be produced by conversion of other energy sources, including fossil fuels and renewables. Only renewables based hydrogen production can contribute to CO_2 emission reduction. Current renewable production methods of hydrogen include H_2 production from biomass, from water by electrolysis (where the electricity has been produced by wind, solar or hydro energy) and the millennium cell alternative, hydrogen on demand.

This company is based in Eatontown, New Jersey, U.S. since 1998 and has a patented process in which a catalysed reaction between water and sodium borohydrate produce hydrogen for applications in cars. The advantage is that the storage of the sodium borohydrate is inherently safe. It is a derivative of borax, which is a natural raw material with substantial natural reserves.

Hydrogen storage

Hydrogen can be stored in different kinds of materials, in gaseous, liquid or more recently in solid form. Gaseous hydrogen can either be transported through natural gas pipelines, mixed into the natural gas, or stored in gas tanks. Liquid hydrogen is stored in metal vessels at high pressures. In solid form, hydrogen is stored in metal hydrides. In the 1980s, the focus shifted to amorphous hydrides such as NiZr and from 1990 the focus is on nanostructured hydrides including carbon nanotubes, nano-magnesium based hydrides, metal hydride-carbon nanocomposites, nanochemical hydrides and alanates. Many companies in U.S. can offer magnesium hydride and sodium sluminium hydride. At the Fraunhofer Institute for Solar Energy in Freiburg, Germany, researchers developed a hydrogen storage device and fuel cell system which is small enough to integrate in a portable digital camcorder.

21.6.4   Cleaner conventional energy

Nanotechnology can also contribute to the improvement of conventional energy sources including coal, oil, gas and nuclear energy and electricity. The report covers both nanotechnology contributions to electricity production and to primary energy production. To start with electricity, the production from coal or natural gas can be made more efficient by using nanotechnology in turbine plants. In nuclear energy, nanotechnology can help improve the radiation resistance of the materials.

21.6.5 Batteries

Batteries are needed to supply electrical energy when you can't get it from the electricity grid. This includes mobile applications such as mobile phones, walkmans, but also home or even village power supply in remote areas and in backup systems in case the grid goes down. In the future, rechargeable batteries will be even more needed in combination with renewable electricity production such as by solar photovoltaics. The sun does not shine when you need the light the most: at night. Even though at the moment both rechargeable and non-rechargeable batteries are available on the market, the trend is towards rechargeables. There are basically two types of rechargeable batteries where nano-structured materials are applied and the focus of research. The first and most advanced is Lithium based, for example Li-ion batteries. These are dry batteries. The other type, wet batteries, uses basically the same materials as for hydrogen storage and are based on metal hydrides, where hydrogen is the chemical energy carrier, or carbon nanotubes. The above mentioned Millenium cell system is also applied in batteries.

21.6.6 Transformation

There are many forms of primary energy, including fossil fuels such as oil and gas, biomass, nuclear energy and renewables such as wind, sun and hydroenergy. These primary energy sources must be transformed into heat, electricity or mechanical power (movement, pressure, etc.). For some of these energy transformations, there is no efficient or cost effective solution. And for some of these needs for new energy transformation technologies, researchers are developing new nanostructured materials or nanocomponents. Fuel cells for transforming hydrogen or other gases (natural gas, methanol) into electricity is a well known example. But researchers are also working on less visible nanotechnologies such as catalysts and membranes for separating different types of gases. These can be used in fuel cells or other energy transforming technologies.

21.6.7 Greening industrial production

A lot of energy is applied in industrial production. This energy can be produced on site for instance by combined heat and power installations, or using the industrial waste as fuel. Industrial production can also contribute to energy saving by using less energy or materials for the same number of products, or by making the products such as cars lighter, hence more energy efficient in their use.

21.6.8 Energy saving

The most sustainable energy use is no energy use. Governments therefore also stimulate energy saving by consumers as well as industry. Some of these measures imply the use of new technologies, such as improved isolation materials. Nanostructured materials such as nanofoams may play a role here.

Thus, nanotechnology research in Europe can contribute to solving future needs for energy technologies, especially in new generations of solar photo-voltaics, the hydrogen economy, more efficient conventional energy production and energy saving for industry as well as consumers. Considering the substantial budgets for research dedicated to nanoresearch including for energy applications, much of this potential is likely to be realised in the coming decades.

21.7   Application of nanotechnology to energy production

Here are some interesting ways that are being explored using nanotechnology to produce more efficient and cost-effective energy:

Generating steam from sunlight: Researchers have demonstrated that sunlight, concentrated on nanoparticles, can produce steam with high energy efficiency. The 'solar steam device' is intended to be used in areas of developing countries without electricity for applications such as purifying water or disinfecting dental instruments. Another research group is developing nano-particles intended to use sunlight to generate steam for use in running power plants.

Producing high efficiency light bulbs: A nano-engineered polymer matrix is used in one style of high efficiency light bulbs. The new bulbs have the advantage of being shatterproof and twice the efficiency of compact fluorescence light bulbs. Other researchers developing high efficiency LED's using arrays of nano-sized structures called plasmonic cavities. Another idea under development is to update incandescent light bulbs by surrounding the conventional filament with crystalline material that converts some of the waste infrared radiation into visible light.

Increasing the electricity generated by windmills: An epoxy containing carbon nanotubes is being used to make windmill blades. Stronger and lower weight blades are made possible by the use of nanotube-filled epoxy. The resulting longer blades increase the amount of electricity generated by each windmill.

Reclaiming waste heat: One of the most promising ways to reduce overall energy use is to recover waste heat from applications such as industrial processes, car engines and electronics and to put this energy to use. Thermoelectric devices, which convert heat gradients directly into electricity, are ideal candidates, but so far their performance has been insufficient for large-scale use. Break throughs in nanotechnology may yield a solution. For example, nanowires made of silicon have a conversion efficiency that is 60 times greater than bulk silicon. Making nanostructured thermoelectric devices out of silicon, which is abundant, cheap and easily handled, could help create a new market for a wide range of devices that recover waste heat. Researchers have used sheets of nanotubes to build thermocells that generate electricity when the sides of the cell are at different temperatures. These nanotube sheets could be wrapped around hot pipes, such as the exhaust pipe of your car, to generate electricity from heat that is usually wasted.

Storing hydrogen for fuel cell powered cars: Researchers have prepared graphene layers to increase the binding energy of hydrogen to the graphene surface in a fuel tank, resulting in a higher amount of hydrogen storage and therefore a lighter weight fuel tank. Other researchers have demonstrated that sodium borohydride nanoparticles can effectively store hydrogen.

Clothing that generates electricity: Researchers have developed piezoelectric nanofibres that are flexible enough to be woven into clothing. The fibres can turn normal motion into electricity to power your cell phone and other mobile electronic devices.

Reducing friction to reduce the energy consumption: Researchers have developed lubricants using inorganic buckyballs that significantly reduced friction.

Reducing power loss in electric transmission wires: Researchers at Rice University are developing wires containing carbon nanotubes that would have significantly lower resistance than the wires currently used in the electric transmission grid. Richard Smalley envisioned the use of nano-technology to radically change the electricity distribution grid. Smalley's concept these upgraded transmission wires, which could transmit electricity thousands of miles with insignificant power losses, with local electricity storage capacity in the form of batteries in each building that could store power for 24 hr use.

Reducing the cost of solar cells: Companies have developed nanotech solar cells that can be manufactured at significantly lower cost than conventional solar cells.

Improving the performance of batteries: Companies are currently developing batteries using nanomaterials. One such battery will be as good as new after sitting on the shelf for decades. Another battery can be recharged significantly faster than conventional batteries.

Improving the efficiency and reducing the cost of fuel cells: Nanotechnology is being used to reduce the cost of catalysts used in fuel cells. These catalysts produce hydrogen ions from fuel such as methanol. Nanotechnology is also being used to improve the efficiency of membranes used in fuel cells to separate hydrogen ions from other gases, such as oxygen.

Making the production of fuels from raw materials more efficient: Nanotechnology can address the shortage of fossil fuels, such as diesel and gasoline, by making the production of fuels from low grade raw materials economical.

Nanotechnology can also be used to increase the mileage of engines and make the production of fuels from normal raw materials more efficient.

Nanotechnology applications having particular relevance to energy transmission technologies: Numerous nonmaterial and other nano-related applications relevant to electricity transmission and petroleum distillate fuel and gas pipeline transport are in various stages of research, development and deployment. These applications have the potential to directly or indirectly reduce the environmental impact associated with the construction, operation and dismantlement of energy transmission technologies.

The remainder of this section highlights examples of nanotechnology applications relevant to transmission of electricity, via, cables and of fossil fuels (i.e., petroleum distillate fuel and natural gas) through pipelines. Potential pitfalls and timeframes have been identified by various researchers. In general, however, the potential for practical scale-up of most of the techniques in use today for nanoparticle production is limited by high capital costs, low production rates, the need for exotic and expensive precursor materials and limited control over nanoparticle physical and chemical homogeneity. Break throughs in nanotechnology research may accelerate the development and implementation of these technologies

Nanotechnology applications relevant to electricity transmission: Nanotechnology may help improve the efficiency of electricity transmission wires. Today, aluminum conductor steel reinforced (ACSR) wire is the standard overhead conductor against which alternatives are compared. Another example, is still in the research phase, is the use of armchair CNTs # a special kind of singlewalled CNT that exhibits extremely high electrical conductivity (more than 10 times greater than copper). Also possessing flexibility, elasticity and tremendous tensile strength, CNTs have the potential, when woven into wires and cables, to provide electricity transmission lines with substantially improved performance over current power lines. Replacing current wires with nanoscale transmission wires, called quantum wires (QWs) or armchair QWs, could revolutionise the electrical grid. The electrical conductivity of QW is higher than that of copper at one-sixth the weight and QW is twice as strong as steel. A grid made up of such transmission wires would have no line losses or resistance, because the electrons would be forced lengthwise through the tube and could not escape out at other angles. Grid properties would be resistant to temperature changes and would have minimal or no sag. (Reduced sag would allow towers to be placed farther apart, reducing footprint and attendant construction and maintenance impacts.) QW, if spun into noncorrosive polypropylene-like rope, could conceivably be buried 'forever' with no fear of corrosion and 'no need for shielding of any kind' Such a grid could have a million times greater capacity than what exists today (assuming the 1-centimeter-diameter aluminum cable carrying about 1000 to 2000 amps); even if the capacity were increased by only 0.1%, the amount of enhanced capacity would still be impressive. The realisation of such conducting possibilities depends on developing processes for producing high-quality CNTs in industrial quantities and at reasonable cost, finding ways to manipulate and orient nanotubes into regular arrays and developing robust testing methods. Today, QWs made from metallic CNTs are very short # no longer than several centimetres # and are manufactured only in limited quantities.

When nanotubes are synthesised, a variety of different configurations appears. Currently, only 2% of all nanotubes can be used as QWs and sorting the armchair nanotubes from the rest is nearly impossible. Current processing technologies are not capable of producing nanotubes with controlled and desirable production properties consistently. Until a good solution for separating the 'good one from the many other unfavourable configurations is reached for large-volume manufacturing, the impact of nanotubes on power line usage is hypothetical'. Long-distance transmission of electrical current entails significant losses (about 20%) due to electrical resistance.

Super conductors transmit electricity with a small fraction of the losses from conventional conductors, thereby enabling power transmission at higher power densities. Such efficiencies may relieve transmission congestion and lessen the need for transmission equipment. High-temperature superconductors (i.e., substances that become superconducting near liquid nitrogen temperatures [about 77 Kelvin (K)] rather than near liquid helium temperatures [about 4 K]) were discovered in the late 1980s. Noting that transmission constraints have contributed to higher electricity prices and reduced reliability, the 2001 National Energy Policy Report (National Energy Policy Development Group 2001) recommended expanded research and development on transmission reliability and superconductivity.

Other electrical transmission infrastructure: Nanotechnology applications may help improve other components of the electric transmission infrastructure, thereby potentially reducing environmental impacts. The examples below pertain to transformers, substations and sensors.

Transformers: Fluids containing nonmaterial could provide more efficient coolants in transformers, possibly reducing the footprints, or even the number, of transformers. Nanoparticles increase heat transfer and solid nanoparticles conduct heat better than liquid. Nanoparticles stay suspended in liquids longer than larger particles and they have a much greater surface area, which is where heat transfer takes place. Using nanoparticles in the development of HTS transformers could result in compact units with no flammable liquids, which could help increase sitting flexibility.

Substations: Substation batteries are important for load levelling peak shaving, providing uninterruptible supplies of electricity to power substation switchgear and for starting backup power systems. Smaller, more efficient batteries could reduce the footprints of substations and possibly the number of substations within a row.

Sensors: Nano electronics have the potential to revolutionise sensors and power-control devices. Nanotechnology-enabled sensors would be self-calibrating and self-diagnosing. They could place trouble calls to technicians whenever problems were predicted or encountered. Such sensors could also allow for the remote monitoring of infrastructure on a real-time basis.

Miniature sensors deployed throughout an entire transmission network could provide access to data and information previously unavailable. The real-time energised status of distribution feeders would speed outage restoration and phase balancing and line loss would be easier to manage, helping to improve the overall operation of the distribution feeder network.

Nanotechnology applications relevant to pipeline transmission of petroleum distillate fuel and natural gas: Today, most of the identified nanotechnology applications for pipelines involve material coatings (insulation, corrosion and multipurpose). Other potential applications include nanosensors, which have the potential to minimise environmental damage by identifying potential leaks before they spread and oil spill remediation with nonmaterial, which may minimise damage should a leak occur. Because the current and expected future applications of nanotechnology for petroleum distillate fuel pipelines are basically the same as those for natural gas pipelines, this section cites examples of general nanotechnology applications for pipelines.

Nanotechnology in energy innovations could impact each part of the value-added chain in the energy sector – energy sources; energy conversion; energy distribution; energy storage and energy usage.

Nanomaterials could lead to energy savings through weight reduction or through optimised function:

1. In the future, novel, nano-technologically optimised materials, for example plastics or metals with Carbon Nanotubes (CNTs), will make airplanes and vehicles lighter and therefore help reduce fuel consumption.
2. Novel lighting materials (OLED: Organic light-emitting diodes) with nanoscale layers of plastic and organic pigments are being developed; their conversion rate from energy to light can apparently reach 50% (compared with traditional light bulbs = 5%).
3. Nanoscale carbon black has been added to modern automobile tyres for some time now to reinforce the material and reduce rolling resistance, which leads to fuel savings of up to 10%.

4. Self-cleaning or 'easy-to-clean' coatings, for example on glass, can help save energy and water in facility cleaning because such surfaces are easier to clean or need not be cleaned so often.

5. Nanotribological wear protection products as fuel or motor oil additives could reduce fuel consumption of vehicles and extend engine life.

6. Nanoparticles as flow agents allow plastics to be melted and cast at lower temperatures.

7. Nanoporous insulating materials in the construction business can help reduce the energy needed to heat and cool buildings.

Nanomaterials could improve energy generation and energy efficiencies:

1. Various nanomaterials can improve the efficiency of photovoltaic facilities.

2. Dye solar cells (Grätzel cells) with nanoscale semiconductor materials mimic natural photosynthesis in green plants.

3. Plastics with carbon nanotubes as coatings on the rotor blades of wind turbines make these lighter and increase the energy yield.

4. Nano optimised lithium-ion batteries have an improved storage capacity as well as an increased lifespan and find use in electric vehicles for example.

5. Fuel cells with nanoscale ceramic materials for energy production require less energy and resources during manufacturing.

6. The effectiveness of catalytic converters in vehicles can be increased by applying catalytically active precious metals in the nanoscale size range.

To sum up to secure global power supply in the long run, it is important not only to develop existing energy sources as efficiently and environmental friendly as possible, but also to minimise energy losses arising during transport from source to end user, to provide and distribute energy for the respective application purpose as flexibly and efficiently as possible and to reduce energy demand in industry and private households.

Nanotechnology provides the potential to enhance energy efficiency across all branches of industry to economically leverage renewable energy production through new technological solutions and optimise the production technologies.

Energy audit in textile industry

22.1   Introduction

An industrial energy audit is a process that facilities energy usage patterns, equipment efficiency and overall building efficiency is determined in order to propose energy efficiency measures. The result of a successful energy audit is decreased energy consumption, reduced raw material usage and increased quality of the end product. The data collected by an energy auditor is the basis on which the energy efficiency suggestions will be created. The implementation of these measures will reduce manufacturing costs and also the negative effects on the environment. Industrial energy audit in other word is a process aimed at finding loopholes in the production process, a design task in order to save raw materials and energy. Performing industrial energy audits makes it possible to save raw materials, energy, optimising the manufacturing process or raising the company's profits and increase competitiveness. After an industrial energy audit, the client data will have an accurate list of energy efficiency measures which will reduce costs and the environmental impact.

22.2   Need for energy audit

In any industry, the three top operating expenses are often found to be energy (both electrical and thermal), labour and materials. If one were to relate to the manageability of the cost or potential cost savings in each of the above components, energy would invariably emerge as a top ranker and thus energy management function constitutes a strategic area for cost reduction. Energy Audit will help to understand more about the ways energy and fuel are used in any industry and help in identifying the areas where waste can occur and where scope for improvement exists.

The energy audit would give a positive orientation to the energy cost reduction, preventive maintenance and quality control programmes which are vital for production and utility activities. Such an audit programme will help to keep focus on variations which occur in the energy costs, availability and reliability of supply of energy, decide on appropriate energy mix, identify energy conservation technologies, retrofit for energy conservation equipment, etc. In general, energy audit is the translation of conservation ideas into realities, by lending technically feasible solutions with economic and other organisational considerations within a specified time frame.

The primary objective of energy audit is to determine ways to reduce energy consumption per unit of product output or to lower operating costs. Energy Audit provides a 'bench-mark' (Reference point) for managing energy in the organisation and also provides the basis for planning a more effective use of energy throughout the organisation.

22.3   Types of energy audits

The type of industrial energy audit conducted depends on the function, size and type of the industry, the depth to which the audit is needed and the potential and magnitude of energy savings and cost reduction desired. Based on these criteria, an industrial energy audit can be classified into two types: A preliminary audit (walk-through audit) and a detailed audit (diagnostic audit).

22.3.1   Preliminary audit (walk-through audit)

In a preliminary energy audit, readily-available data are mostly used for a simple analysis of energy use and performance of the plant. This type of audit does not require a lot of measurement and data collection. These audits take a relatively short time and the results are more general, providing common opportunities for energy efficiency. The economic analysis is typically limited to calculation of the simple payback period, or the time required paying back the initial capital investment through realised energy savings.

22.3.2   Conducting the preliminary analysis

The preliminary analysis helps the energy auditor to better understand the plant by providing a general picture of the plant energy use, operation and energy losses. This effort provides enough information to undertake any necessary changes in the audit plan. In the preliminary analysis, a flowchart can be constructed that shows the energy flows of the system being audited (Fig. 22.1). An overview of unit operations, important process steps, areas of material and energy use and sources of waste generation are presented in this flowchart. The auditor should identify the various inputs and outputs at each process step. The preliminary flowchart is simple, but detailed information and data about the input and output streams can be added later after the detailed energy audit. An example of a preliminary flowchart for a textile dying plant is shown in Fig. 22.2.

22.3.3   Detailed audit (diagnostic audit)

For detailed (or diagnostic) energy audits, more detailed data and information are required. Measurements and a data inventory are usually conducted and different energy systems (pump, fan, compressed air, steam, process heating, etc.),

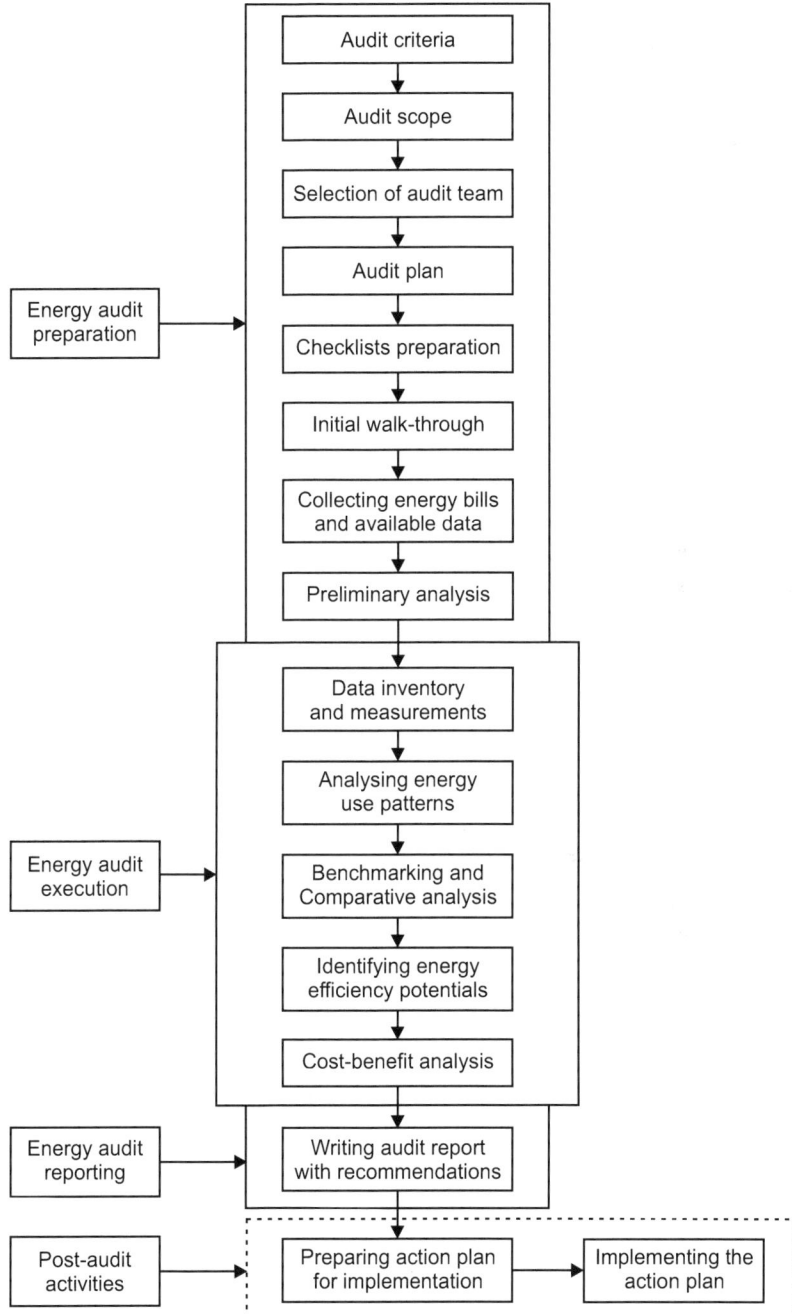

Figure 22.1: Overview of an industrial energy audit.

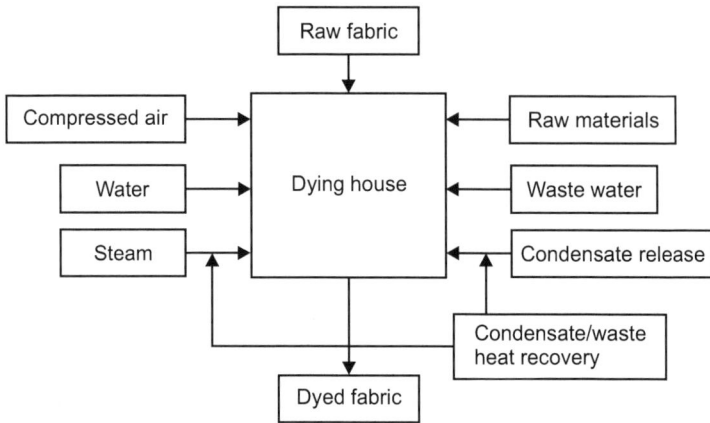

Figure 22.2: Energy flowchart for a textile dying plant.

are assessed in detail. Hence, the time required for this type of audit is longer than that of preliminary audits. The results of these audits are more comprehensive and useful since they give a more accurate picture of the energy performance of the plant and more specific recommendation for improvements. The economic analysis conducted for the efficiency measures recommended typically go beyond the simple payback period and usually include the calculation of an internal rate of return (IRR), net present value (NPV) and often also life cycle cost (LCC). These three main steps are: energy audit preparation, execution and reporting.

22.3.4   Conducting detailed energy audit

A detailed energy audit provides a comprehensive energy project implementation plan for a facility, since it evaluates all major energy-using systems.

This type of audit offers the most accurate estimate of energy savings and cost. It considers the interactive effects of all projects, accounts for the energy use of all major equipment and includes detailed energy cost saving calculations and project cost.

In a detailed audit, one of the key elements is the energy balance. This is based on an inventory of energy-using systems, assumptions of current operating conditions and calculations of energy use.

This estimated use is then compared to utility bill charges.

Detailed energy auditing is carried out in three phases:

1. Phase I – Pre-audit.
2. Phase II – Audit.
3. Phase III – Post-audit.

22.4 Analysing energy bills and inventory measurement of energy use

22.4.1 Analysing energy bills

Energy bills, especially those for electricity and natural gas, are very useful for understanding and analysing a plant's energy costs. It is important to understand the different components of these bills, so that a correct and helpful analysis can be conducted.

Electricity bills

Several costs are usually included in the electricity bill. Most electric rates include a fixed service (or customer) charge that is constant regardless of the amount of electricity used and a per kilowatt-hour (kWh) rate for the amount of electricity consumed. Electricity use in the period covered by the electricity bill can be divided by the number of the days given in the bill. Since reading periods in the bills can vary, kWh/day is more useful for identifying consumption trends than the total billed kWh. This can be used later to accurately calculate the monthly electricity use and can also be used for graphical analysis.

Calculating the load factor (LF): The load factor is the ratio of the energy consumed during a given period (in the electricity bill) to the energy which would have been consumed if maximum demand had been maintained throughout the period.

22.4.2 Inventory and measurement of energy use

Gathering data through an inventory and measurement is one of the main activities of energy auditing. Without adequate and accurate data, an energy audit cannot be successfully accomplished. Some data are readily available and can be collected from different divisions of the plant being audited. Some other data can be collected through measurement and recording. The energy audit team should be well-equipped with all of the necessary measurement instruments. These instruments can be portable or installed in certain equipment. The most common data measured during the auditing process are:

1. Liquid and gas fuel flows.
2. Electrical measurements, such as the voltage, current intensity and power, as well as power factor.
3. Temperatures of solid and liquid surfaces.
4. Pressure of fluids in pipes, furnaces or vessels.
5. Exhaust gases emissions (CO_2, CO, O_2 and smoke).
6. Relative humidity.
7. Luminance levels.

Electrical load inventory

Making an inventory of all electrical loads in a plant aims to answer two important questions: where the electricity is used? How much and how fast is electricity used in each category of load? One way to prioritise the electricity-saving opportunities is by the magnitude of the loads. Therefore, identifying and categorising different loads in a plant can be useful.

Thermal energy use inventory

An energy flow diagram like the one shown in Fig. 22.2 is helpful for identifying thermal energy flows. The energy flow chart can show all energy flows into the facility, all outgoing flows from the facility to the environment and all significant energy flows within the facility.

The purpose of an energy flow diagram is not to describe a process in detail. In fact, it will generally not show specific devices and equipment that are found in its various sub-systems. The sum of the energy outflows should equal energy inflows. With this information, it is often possible to see opportunities for energy saving and recovery.

22.4.3 Benchmarking and comparative energy performance analysis

Energy efficiency benchmarking and comparisons can be used to assess a company's performance relative to that of its competitors or its own performance in the past. Benchmarking can also be used for assessing the energy performance improvement achieved by the implementation of energy-efficiency measures. Also, on a national level, policy makers can use bench-marking to prioritise energy-saving options and to design policies to reduce greenhouse gas emissions. International comparisons of energy efficiency can provide a benchmark against which a company's or industry's performance can be measured to that of the same type of company or industry in other countries.

Benchmarking energy performance of a facility enables energy auditors and managers to identify best practices that can be replicated. It establishes reference points for managers for measuring and rewarding good performance. It identifies high-performing facilities for recognition and prioritises poor performing facilities for immediate improvement.

While conducting benchmarking, the key drivers of energy use should be identified and the benchmarking metrics might be adjusted or normalised, for instance, based on the weather, production levels, or product characteristics that affect energy use. Successful benchmarking also requires monitoring and verification methods to ensure continuous improvement.

22.5   Energy savings and audit in textile industry

22.5.1   Major areas of energy saving potential

Ring frame parameters

The spindle consumes 30% of power consumed by ring frame. The optimum spindle speed is the ultimate parameter in the mill, which we are targeting to achieve. The optimum spindle speed is the speed @ which ring frame gives more output speed with less power input and keeping the prime mover and transmission (by flat belts, etc., which minimises linkage losses) under healthy condition. To improve ring frame performance, after looking into the textile-associated savings, let us consider motor and linkage parameters. Condition based monitoring of the following parameters will definitely enhance its health and productivity and minimising energy costs.

Rewound motor efficiency: Rewound motors are working at reduced efficiency. The Efficiency Bell curve indicates peak efficiency of a standard motor at its three quarter loading. Here in rewound motor, the efficiency peaks at the lower loading level only. And the slippage in the motor increases nearing its full load. Hence keep an eye on these motors thermal characteristic, we must not fully load the motor, but match our process to its reduced efficiency.

Measure parameters: Measure motor parameters daily.

Improving motor efficiency: We have to take care of the motor with positive active ventilation all over. For the same, we ensure strong axial air throw along the ribs of motor so that overall surrounding temperature of the motor comes down. Now the motor breathes normally with the shrouded fan effect at its one end and its efficiency improvement is seen in the long run. The loss to the motor due to this retrofit is very minimal, but overall efficiency of the motor improves due to the above force cooling of fins and the motor itself.

Efficient humidification

1. Textile mill stalwarts have understood now that by running HP (humidification plant) scientifically they can improve production. To be precise, if we concentrate on HP, precisely at the Spray Dwell Time of humidified air in the spray chamber, then we can improve the HP performance and this definitely gives a boost in output yarn produced.

2. Usage of high efficiency FRP fans (properly sized to the air throw and cfm specs and installed correctly) instead of M.S. or Aluminium blades consume less power input for the given air output.

3. Spot capacitor @ the motor terminals inside HP premises is more important and this aspect is often neglected in many mills.

Water audit

1. For humidification plant, water softening plant is the first priority. Frequent Water analysis is a must before putting the water to use inside the equipments so that we know the TDS, pH, etc., parameters of the incoming water and the used water; this will help to ascertain the scaling effects in the wetted parts of the equipments and pipe lines, etc.

2. In boiler, prevention of feed water scaling is cost effective compared to blow down losses considering overall boiler's health. Similarly the industry has still now accepted to use whatever water it receives. Giving forethought to wetted parts of the equipments in the long run future usage, we must give only softened water as water input.

3. Rainwater harvesting done recently in mill premises definitely helped to improve the condition of existing water to improve its hardness, etc.

4. The water pump is being downsized to half in mills now. To increase the spray dwell time in the spray chamber, the pump discharge is around 2 kgsc. Provision must be made to measure the water pressure at pump discharge and at the spray header end after the spray nozzles, whether it is clamp on PVC nozzle or gunmetal bulk nozzle. Nozzle must spread water mist at low pressure than jet out water at high pressure.

5. These pressures are indicative of correct atomisation at the rated pressure of nozzles and show us the shoot up in pressure in case of choke in nozzles. Let us first check whether we are pumping water at the rated pressure of nozzle regularly.

6. The temperature of the cooling water sump will indicate (on continuous monitoring) the water circuit resistance in the air washer area and its choking status and the need to clean up the air washers.

Compressed air system

1. Compressor is like a submersible pump working in air.

2. Compressor needs air to suck in and to surround, to deliver more air.

3. Cool compressor delivers more air.

4. Continuously run compressor delivers less than an intermittently run Compressor.

5. Tune your compressor cut-in/cut-out to deliver more cool air.

The air compressor in the utility in any plant is similar to submersible pump in functioning. Water is the sucking medium for the pump inside and same water is cooling medium outside. Similarly air compressor starves if its air filter is choked and as well compressor-surrounding temperature gets heated up locally. We are aware that in the boiler it is relatively cost effective energy saving to soften the feed water and increase its temperature compared to the post combustion methods. Similarly it is cost effective to ensure cool and correct volume of air is ensured at the air compressor suction than to strain more to reduce the discharge heat and reduce wear and tear of internals.

22.6   Understanding energy costs

Understanding energy cost is vital factor for awareness creation and saving calculation. In many industries sufficient meters may not be available to measure all the energy used. In such cases, invoices for fuels and electricity will be useful. The annual company balance sheet is the other sources where fuel cost and power are given with production related information.

Energy invoices can be used for the following purposes:

1. They provide a record of energy purchased in a given year, which gives a base-line for future reference.
2. Energy invoices may indicate the potential for savings when related to production requirements or to air conditioning requirements/space heating, etc.
3. When electricity is purchased on the basis of maximum demand tariff.
4. They can suggest where savings are most likely to be made.
5. In later years invoices can be used to quantify the energy and cost savings made through energy conservation measures.

22.7   Cost-benefit analysis of energy-efficiency opportunities

After identifying the list of energy-efficiency measures applicable to the facility, the auditor can also conduct an economic feasibility analysis, a so-called 'cost-benefit analysis', for the measures and make recommendations for their implementation. Step-by-step guidance for energy auditors on life-cycle costing, discounting, net present value, internal rate of return, savings-to-investment ratio and payback periods in order to conduct the common economic analysis for the assessment of financial viability of the energy efficiency measures are discussed below.

22.7.1   Life-cycle cost analysis (LCCA)

Life-cycle cost analysis (LCCA) is an economic method of project financial evaluation in which all costs from owning, operating, maintaining and disposing of a project are taken into account. LCCA is useful for evaluating energy-efficiency projects because the capital cost of energy-efficiency projects is incurred at once at the beginning of the project, while the savings occur throughout the lifetime of the project. Hence, LCCA can determine whether or not these projects are economical from the investor's viewpoint, based on reduced energy costs and other cost reductions over the project or equipment lifetime.

Also, there are often a number of cost-effective alternatives for energy-efficiency improvement of the system. In such cases, LCCA can be used to identify the most cost-effective alternative for a given application. This is normally the alternative with the lowest lifecycle cost. LCCA stands in direct contrast to the simple payback period (SPP) method which focuses on how quickly the initial investment can be recovered. As such SPP is not a measure of long-term economic performance or profitability or the project. The SPP method typically ignores all costs and savings occurring after the point in time in which payback is reached. It also does not differentiate between project alternatives having different lifetimes and it often uses an arbitrary payback threshold. Moreover, the SPP method which is commonly used ignores the time value of money when comparing the future stream of savings against the initial investment cost.

22.7.2 Life cycle cost (LCC) method

The LCC is the total cost of owning, operating, maintaining and disposing of the technology over the lifetime of the project or technology. In this method, all costs are adjusted (discounted) to reflect the time value of money. The LCC of a technology or measure has little value by itself; it is most useful when it is compared to the LCC of other alternatives which can perform the same function in order to determine which alternative is most cost effective for this purpose. These alternatives are typically 'mutually exclusive' alternatives because only one alternative for each system evaluated can be selected for implementation.

22.8   Summary for conducting detailed energy audit

A 10-step summary for conducting a detailed energy audit at the field level is listed in Table 22.1.

Table 22.1: 10 Steps for a detailed energy audit.

Step	Action	Purpose
1	*Phase I – pre-audit* • Plan and organise • Walk-through audit • Informal interviews with energy manager, production/plant manager	• Resource planning; establish/organise energy audit team • Organise instrumentation and time frame • Macro data collection (suitable to type of industry)

(Cont'd...)

Step	Action	Purpose
		• Familiarisation of process/plant activities • First-hand observation and assessment of current level operation and practices
2	Conduct briefing/awareness session with all divisional heads and persons concerned (2–3 hrs)	• Building up cooperation • Issue questionnaire for each department • Orientation, awareness creation
3	*Phase II – audit* Primary data gathering, process flow diagram and energy utility diagram	Historic data analysis; baseline data collection Prepare process flowchart(s) All service utilities system diagram (Example: Single line power distribution diagram, water, compressed air and steam distribution) Design, operating data and schedule of operation Annual energy bill and energy consumption pattern (refer to manuals, log sheets, equipment spec. sheets, interviews)
4	Conduct survey and monitoring	Measurements: • Motor survey, insulation and lighting survey with portable instruments to collect more and accurate data • Confirm and compare actual operating data with design data
5	Conduct detailed trials/experiments for biggest energy users/equipment	Trials/experiments: • 24-hr power monitoring (MD, PF, kWh, etc.) • Load variation trends in pumps, fans, compressors, heaters, etc. • Boiler efficiency trials (4–8 hrs) • Furnace efficiency trials • Equipment performance experiments, etc.
6	Analysis of energy use	Energy and material balance and energy loss/waste analysis
7	Identification and development of energy conservation (ENCON) opportunities	Identification and consolidation of ENCON measures Conceive, develop and refine ideas Review ideas suggested by unit personnel Review ideas suggested by preliminary energy audit

(Cont'd...)

Step	Action	Purpose
		Use brainstorming and value analysis techniques
		Contact vendors for new/efficient technology
8	Cost-benefit analysis	Assess technical feasibility, economic viability and prioritisation of ENCON options for implementation
		Select the most promising projects
		Prioritise by low-, medium-, long-term measures
		Documentation, report presentation to top management
9	Reporting and presenting to top management	Documentation, report presentation to top management
10	*Phase III – post-audit*	
	Implementation and follow-up	Assist and implement ENCON measures and monitor performance
		Action plan, schedule for implementation
		Follow-up and periodic review

22.9 Preparing an energy audit report

After finishing the energy audit, the audit team should write an energy audit report. In the report, the auditors should explain their work and the results in a well-structured format. The energy audit report should be concise and precise and should be written in a way that is easy for the target audience to comprehend. Some key issues that should be kept in mind while writing an audit report are:

1. The audit report should be written in a way that provides suitable information to the potential readers of the report which could be the CEO or plant manager, the supervisor of engineering or maintenance and the plant shift supervisor.

2. The audit report should be concise and precise and use direct language that is easy to understand.

3. Use more graphs rather than tables for the presentation of data, results and trends.

4. The recommendation section should be specific, clear and with adequate detail.

5. Assumptions made in the analysis should be explained clearly. How changes in the key assumptions can influence the results should also be explained. A sensitivity analysis is a very helpful tool for this.

6. The auditors should do their best to avoid mistakes and errors in the report especially in the results. Even a few errors could damage the credibility of the audit.

7. The energy audit report should be consistent in structure and terminology used.

8. The calculations made in the analysis work should be explained clearly. An example of each type of calculation can be given either in the main body of the report or in appendix for more clarity.

Typical energy audit report contents and format are given below. The following format is applicable for thorough energy audit of a plant in most industries. However, the format can be modified for specific requirements applicable to a particular type of industry or energy audit.

22.9.1   Typical content for an energy audit report

1. Executive summary:
   - (a) Summary information on key audit findings (annual consumption and/or energy budget, key performance indicators, etc.).
   - (b) Recommended energy-efficiency measures (with a brief explanation of each).
   - (c) Implementation costs, savings and economic indicators (e.g., IRR, NPV, SPP) for the recommended measures.
   - (d) Any other useful information related to the implementation of energy-efficiency measures.

2. Audit objectives, scope and methodology.

3. Plant overview:
   - (a) General plant details and description.
   - (b) Component of production cost (raw materials, energy, chemicals, manpower, overhead, other).
   - (c) Major energy use and the users.

4. Production process description:
   - (a) Brief description of manufacturing process.
   - (b) Process flow diagram.
   - (c) Major raw material inputs, quantities and costs.

5. Energy and utility system description:
   - (a) List of utilities.
   - (b) Brief description of each utility (any of the following that are applicable).
     - Electricity.

- Steam.
- Water.
- Compressed air.
- Chilled water.
- Cooling water.
- Process heating.

6. Detailed process flow diagram and energy and material balance:
   (a) Flowchart showing flow rate, temperature and pressures of all input-output streams.
   (b) Water balance for major units in the facility.

7. Energy use analysis in utility and process systems (any of the following that are applicable):
   (a) Boiler efficiency assessment.
   (b) Furnace/kiln efficiency analysis.
   (c) Cooling water system performance assessment.
   (d) Refrigeration system performance.
   (e) Compressed air system performance.
   (f) Summary of load inventory results.

8. Energy use and energy cost analysis in the plant:
   (a) Specific energy consumption.
   (b) Summary of energy bills analysis results.
   (c) Summary of the results from the analysing the energy use and production patterns.
   (d) Summary of the results from the benchmarking analysis.
   (e) Summary of assumptions and samples for all important calculations.

9. Energy-efficiency options and recommendations:
   (a) List of energy-efficiency options classified in terms of no cost/low cost, medium cost and high investment cost along with their annual energy and cost savings.
   (b) Summary of the cost-benefit analysis of energy-efficiency measures.

10. Conclusion and a brief action plan for the implementation of energy-efficiency options.

References

Adams Douglas, *Principles of Energy Conversion*, Academic Press, New York.

Agatha Christie, *Energy Conservation*, by Cavendish Square Publishing.

Bauer, E., *Energy Conservation Equipments*, Van Nostrand Reinhold, USA.

Beason Doug, *Industrial Heat Recovery*, Interscience, New York.

Charles Wing, *Visual Handbook of Energy Conservation*, Pergamon Press.

Cherryh, C. J., *Handbook of Energy Conservation*, Taunton Press, UK.

Dale, R. Patrick, *Energy Conservation Guidebook*, Oxford Press, London.

David, A. Reay, *Handbook of Industrial Energy Conservation*, Reinhold, USA.

Denise Chesterton, *Industrial Energy Conservation*, Pergamon Press.

Donald, R. Wulfinghoff, *Energy Efficiency Manual*, Energy Institute Press, New York.

Fardo, W.S., *Pumps and Motors*, Cambridge University Press, Cambridge.

Gaddis William, *Heat Recovery Equipments*, Henser, USA.

Garrett Randall, *Waste Heat Recovery*, Van Nostrand Reinhold, USA.

Jacques Brian, *Energy Options: Challenge for the Future*, Noyes, UK.

James, M.R., *Energy Management*, Noyes, UK.

John, S. Rinaldi, *Energy Techniques for Industries*, Van Nostrand, Reinhold, USA.

Katzenbach John, *Elements of Energy Conversion*, Reinhold, USA.

Mikhail, J. Ivanov, *Energy Conservation Techniques*, Academic Press, New York.

Morton, M.J, *Energy Conservation Techniques*, Van Nostrand, Reinhold, USA.

Palmer, R.A., *Industrial Energy Conservation*, Maclaren, London.

Patrick, R.D., *Energy Conservation in Industrial Sector*, University of Wisconsin Press, Madison.

Paul, W. O'Callaghan, *Handbook of Energy Conservation*, Pergamon Press,

Penn, W.S., *Industrial Energy Conservation*, Maclaren, London.

Revonna, M. Bieber, *The Science of Renewable Energy*, Maclaren, London.

Richardson, E.R., *Energy Conservation in Textile Industries,* Humana Press Inc., New Jersey.

Roger, W.R., *Energy Efficiency by Audit,* Academic Press, New York.

Saberhagen Fred, *Direct Energy Conversion*, Noyes, UK.

Saroyan William, *Energy Costs in Small Business and Industries*, Reinhold, USA.

Solmes, Leslie A., *Energy Efficiency*, Springer Netherlands.

Steve, D.T., *Energy Management Handbook*, by Fairmont Press, New York

Turner, W.C., *Boiler and Turbines,* Ellis Harwood, New York.

Vargic Martin, *Developments in Heat Transfer,* Elsevier, USA.

Wallace Bronwen, *Handbook of Fuel Cell Technology,* Maclaren, London.

Washington, T. Booker, *Energy Conversion and Utilisation*, Mcgraw-Hill, New York.

Index